CAMBRIDGE COMPARATIVE PHYSIOLOGY

GENERAL EDITORS:

J. BARCROFT, C.B.E., M.A.
Fellow of King's College and ,Professor of
Physiology in the University of Cambridge

and

J. T. SAUNDERS, M.A.
Fellow of Christ's College and Demonstrator
of Animal Morphology in the University of
Cambridge

THE
GENETICS OF SEXUALITY
IN ANIMALS

THE
GENETICS OF SEXUALITY
IN ANIMALS

BY

F. A. E. CREW, M.D., D.Sc., Ph.D.

*Director of the Animal Breeding Research Department and Lecturer
in Genetics in the University of Edinburgh*

CAMBRIDGE

AT THE UNIVERSITY PRESS

1927

CAMBRIDGE
UNIVERSITY PRESS

University Printing House, Cambridge CB2 8BS, United Kingdom

Cambridge University Press is part of the University of Cambridge.

It furthers the University's mission by disseminating knowledge in the pursuit of education, learning and research at the highest international levels of excellence.

www.cambridge.org
Information on this title: www.cambridge.org/9781107502505

© Cambridge University Press 1927

First published 1927
First paperback edition 2015

A catalogue record for this publication is available from the British Library

ISBN 978-1-107-50250-5 Paperback

AUTHOR'S PREFACE

It is not without significance that a geneticist should be asked to make a contribution to a series of monographs on Comparative Physiology. It seems reasonable to interpret this as a recognition on the part of the physiologist of the desirability of including genetics in any comprehensive survey of general physiology.

The science of genetics has hitherto been mainly concerned with the significance of ratios between classes of related individuals which appear in successive generations of an experiment starting with two types that differ from one another in respect of one or more heritable characters. It has defined the germplasm, "mapped" the chromosomes, and has demonstrated the intimate relation between somatic character and some definite antecedent determiner in the gamete. But so busy has the geneticist been collecting and interpreting his ratios, and so inefficient are the present day techniques of developmental physiology, that as yet very little effort has been made to demonstrate the nature of this relationship. However, the geneticist has always recognised that the facts which he has disclosed demand a physiological interpretation, that genetic phenomena must be accepted as evidence of the action of a long chain of physiological processes during development and differentiation.

It is reasonable to hold that in all probability the present methods of genetics, so ably used, have already made their great and lasting contributions to biological science. There are still many gaps to be filled and many extensions and applications to be made, but the great general principles that could emerge from the purely genetical mode of investigation have already been disclosed. Formal genetics as a tool wherewith to explore the great problem of evolution is becoming blunted. At least, so it seems to me.

I am very conscious of the shortcomings of my treatment of the subject. Since the number of pages was strictly limited and the subject is extensive and ever enlarging, I have been forced to omit or to treat inadequately much that is of importance. I dread lest in my presentation I may chill the interest in my subject that

I would awaken in the reader. I hope that may not be so. I was encouraged to write this book by the hope that in the reading of it someone may be sufficiently stirred as to decide to devote himself to a subject which has always fascinated me. To him I would say that there is no more fruitful field of work, and none more wonderful, and that the experimental biologist who wishes to serve his science would be well advised to consider specialising in developmental physiology and devoting himself to the problem as to how the gene in its action produces its end result, the character.

F. A. E. C.

1927

CONTENTS

Chapter I

THE MECHANISM OF SEX-DETERMINATION

1. INTRODUCTORY.

Sex is the term used to define that differentiation of different parts of an individual, or of the same individual at different times, or of different individuals, which is associated with the elaboration of physiologically and often morphologically dissimilar gametes in the union of which the next generation of individuals has its origin. Maleness is the state or quality associated with the elaboration of spermatozoa (or of their physiological equivalents); femaleness is the state or quality associated with the elaboration of ova (or of their physiological equivalents). Sexuality is the state or quality of being distinguished by sex. A male is an individual that exhibits the state or quality of maleness, one that is efficiently equipped for the elaboration of functional spermatozoa and for the conveyance of these towards the site of fertilisation; a female, one that exhibits the state or quality of femaleness, one efficiently organised for the elaboration of functional ova, for the conveyance of these to the site of fertilisation, and often for the prenatal accommodation of the zygote, the fertilised egg, for the transit to the exterior of this new individual at some stage of its development and for the nurture of it thereafter. If in a group (*e.g.* a species) it is customary for maleness and femaleness to be exhibited by one and the same individual, coincidently or in succession, the group and the individuals comprising it are monoecious, hermaphroditic, though it follows that in certain cases an individual can at one time be a male and at another a female. If in a group it is not customary for maleness and femaleness to characterise one and the same individual the group is dioecious, bisexual; the sexes are separate, and every individual within the group is throughout its sexual life either a male or else a female.

Sexuality is an attribute of the function of reproduction; it is concerned with the capacity of living things to multiply. Sexual reproduction (amphigony), distinguished by the preliminary process of fertilisation (syngamy), requires that two physiologically and in many instances morphologically dissimilar gametes derived, in most cases, from physiologically dissimilar areas of one and the same individual or from two separate and sexually distinct individuals, shall unite to form a zygote in which the new individual shall have its beginning.

Sexual reproduction has been demonstrated in all groups of the non-cellular Protozoa. When the different modes of such sexual reproduction are compared with that which obtains in the case of the Metazoa, it is found that certain essential features are common to both. There is in both cases reduction of nuclear material and, following this, a fusion of nuclear material derived from different sources.

Sexual reproduction is practised by all forms of animal life and it is reasonable to assume that it is beneficial to the race. It would appear that the reproductive elements of the individual body are not involved in the general bodily functioning, in the processes of individuation, but lie dormant and protected within the body, specially reserved for their own particular destiny. They constitute the material chain that binds the generations and are in a sense immortal. It can be shown in favourable instances that this segregation of the reproductive elements is a fact, and it is probable that it is so in all groups of organisms. In *Ascaris megalocephala* it has been possible to demonstrate that the first division of the fertilised egg results in two cells which can be recognised from the beginning as being different in their organisation; one of these cells gives rise to the somatic tissues and the other to the germ-cells. In the case of the former the nuclear material undergoes a marked diminution in quantity whereas in the other no such reduction takes place. The lineage of the sex-cells from embryo to adult is demonstrable and these can be shown to form the material link between the generations.

Bisexuality can be regarded as the basis of evolutionary plasticity since through it mutations that have occurred remotely in space and independently in time can meet to reinforce each other, offering among the number of their random combinations characterisations which, when tested by the selective agencies within their environment, are judged on their merits, the harmful leading to the elimination of their exhibitors. The possible benefits of bisexuality seem to be foreshadowed in some degree in the case of the conjugation of *Paramoecium*. Jennings (1913), in the course of extended and critical experimentation, has demonstrated perfectly clearly that conjugation is not in itself necessarily beneficial and that its value lies in the opportunity it provides for the attainment of the state of heterozygosis. For an explanation of this reference must be made to the Chromosome Theory of Heredity. If it is true, as this hypothesis postulates, that for all the characters, anatomical and physiological, there are antecedent determiners, factors, or genes, in the germplasm, the chromosomes themselves; if new characterisations are but reflections of specific regional alterations in the organisation of the chromatin material, of mutations, and if new characters, having arisen, persist in virtue of the integrity of the hereditary mechanism, then it follows that in allogamy—cross-fertilisation—there exists the mechanism for the spread of a new characterisation through the race to which the individual belongs, since it can be brought into association with other genetic variations, other genetic deviations from the usual characterisation, that have occurred independently in time and remotely in space. This mingling of different hereditary constitutions, of different genotypes, leads to different factorial recombinations, and thence to new character combinations, new phenotypes, these being the raw material upon which selective agencies may work. Thus if N mutations occur in the germplasm of an asexually reproducing organism, only N phenotypes can arise, whereas if N mutations occur in the germplasm of a sexually reproducing organism, 2^N phenotypes can be formed. Ten mutations mean 10 phenotypes in the first case, 1024 in the second. If for every heritable character there is an antecedent determiner or gene, then this gene may be present in the germplasm in the duplex state, having been contributed by both parents, or it may be present in the simplex state, having been contributed by one. All characters are not advantageous; the harmful can, however, be balanced by the helpful. The end-result of a gene in the simplex state may not be so disadvantageous as that of the same gene in the duplex state. A heterozygous individual, possessing the gene (or genes) for a character (or characters) in the simplex state may thus be better fitted to accommodate itself to the variations of an inconstant environment than the homozygote, which, because of the purity of its factorial constitution, is not so plastic. By means of conjugation and of allogamy generally advantageous genetic acquisitions can be pooled. Equally truly, disadvantageous genetic acquisitions can be pooled also, but if it is permissible to speak of advantage at all, it is to the advantage of the race and not of the individual that reference is made.

2. SEX-DIMORPHISM.

In the great majority of animals every individual is either a male or else a female. Male is commonly to be distinguished from female by differences in the sexual phenotype, the sexual characterisation consisting of (1) the gonads or sex glands, (2) the accessory sexual apparatus of ducts and associated glands concerned with the transference of the products of the gonads and in the female of many forms of the zygote itself, (3) the external organs of reproduction, and (4) certain skeletal and cutaneous and other less definite physiological and psychological characters, often loosely referred to as the secondary and tertiary sexual characters, some of which are employed not directly in sexual congress but in some cases in courtship, combat, concealment, and in the case of the female, in the care and nourishment of the young[1]. Gräfenberg (1922), Manoilov (1922–3), Satina and Demerec (1925), and Edlbacher and Röthler (1925), amongst others, have recently described certain pieces of experimentation that bear directly upon this question of sex-dimorphism for they confirm the impression that the sexes are biochemically distinct.

The relegation of the function of reproduction to a specialised system of the individual's body has been attended by the provision of an efficient equipment for sexual congress. In the simpler forms the products of the gonads are merely liberated at the body surface and the prospects of the fertilisation of the ovum by the sperm are relatively remote; fertilisation is entirely a matter of the chance

[1] Geddes and Thompson, 1899; Cunningham, 1900; J. S. Huxley, 1923; Havelock Ellis, 1914; Morgan, 1914; Meisenheimer, 1922; and Goldschmidt, 1923; treat this question of sex-dimorphism fully.

meeting of dissimilar gametes. The development of the accessory sexual apparatus and external genitalia provided the means for the direct transference of spermatozoa to the genital passages of the female and so rendered fertilisation far more certain. Highly elaborated sexual tropisms, when developed, further increase the certainty of profitable sexual congress, whilst amazingly perfect contraptions, such as the phosphorescent organ of the firefly, *Photinus pyralis*, have been utilised to bring the sexes together for this purpose. Care of the young is exhibited by those species in which the production of gametes, particularly by the female, is restricted, and in those cases in which the offspring themselves possess but a limited supply of nutritive material. As a general rule, it seems that the earlier an embryo is forced by its organisation actively to provide for its own subsistence, the more care it needs. The mother in some cases merely protects her offspring, in others she also nurses them. In some cases the young actively seek parental protection, in others the father and not the mother cares for the embryonic young, *e.g. Alytes obstetricans*. The parents in some cases carry the developing zygotes in different parts of the body which often take the form of specialised sacs, *e.g.* the marsupial pouch and the uterus. Oviparity is replaced by viviparity with associated changes in the maternal organism, since it demands internal fertilisation and the anatomical, physiological, and psychological equipment for coitus. Such are the evidences of *storge*, the postulated hereditary instinct which tends to preserve a species.

Though attempts to classify the sex-dimorphic characters for purposes of discussion have been many, it has to be confessed that as yet no satisfactory classification exists[1]. That this is so is due to the fact that as yet no exact knowledge exists concerning the genetic nature of many of these characters and of their relation to the sex-glands, and until this has been secured, any attempt to classify them must be premature. For the present, it is enough to hold that the characters which in their combination constitute maleness or femaleness respectively are characters in the genetic sense, being expressions of a genotype that is determined by the nature of the genes brought into the zygote by the conjugating gametes. In the

[1] See Lipschütz, 1924; Zavadovsky, 1926; Baker, 1926.

fertilised egg there are none of the characters that distinguish race from race, male from female, that make each individual the first and last of its identical kind; these are expressed as ontogeny proceeds and their expression is modelled, encouraged, embarrassed, to a greater or lesser extent, by the impress of the agencies of the external environment, the outer world, and by the changing conditions within the developing zygote itself, the internal environment, established by the whole of the genotype in action and by the functional activity of the characters as these become differentiated. As a working hypothesis, it can be accepted that sexuality is primarily based upon antecedent determiners within the germplasm, upon genes resident in the chromosomes.

3. THE TIME WHEN THE SEX OF THE INDIVIDUAL
IS DETERMINED AND THE MECHANISM BY
WHICH IT IS DETERMINED.

THE EVIDENCE OF POLYEMBRYONY.

At the beginning of this century, it was generally believed that at the time of fertilisation the egg was completely ambivalent as regards the future sex of the resulting zygote; it was customary to refer the sex of an organism to the conditions incident to development. But certain facts of general biology are now known which are not susceptible to interpretation of this kind. They point to the view that sex in the higher animals is usually predetermined at the time of fertilisation. Identical twins, *i.e.* twin zygotes derived from a single fertilised ovum, are always of the same sex. Such polyembryony is rare in the human, but in the Texas nine-banded armadillo it is the rule for four young to be produced at a time, all of the same sex and remarkably alike. Newman and Paterson (1909) were able to show that in the case of the armadillo a single fertilised ovum after development to a certain stage budded off four embryos. On the other hand, in those cases in which the different embryos arise from separate ova, it is known that the individuals are not invariably of the same sex. There is no appreciable reason why, if purely environmental factors are at work in determining the sex of the offspring, litters produced from one egg should be of the same sex, while litters produced from separate eggs should include both males and females. From such observations as these it would appear that the sex of the individual is determined by the constitution of the fertilised ovum at the time of fertilisation.

CYTOLOGICAL EVIDENCE.

In an ever increasing number of instances it can be shown that the phenotypic differences distinguishing male and female are associated with constant differences in the chromosome content of the tissues of the two sexes. This fact is illustrated in the clearest possible manner in the case of *Drosophila melanogaster* which has four pairs of homologous chromosomes. In the somatic cells and immature gametes of the male one of the four pairs is remarkable in that its members are morphologically dissimilar, whereas in these tissues of the female the members of this pair are, as is the case in all the other pairs in both male and female, morphologically alike. Since tissues from male and female differ chromosomally only in this respect, these chromosomes are referred to as the sex-chromosomes, and the members of the sex-chromosome pair in the female and the one sex-chromosome of the male that is morphologically similar to these are known as the X-chromosomes, whilst the unequal mate of the X in the male is known as the Y-chromosome. In respect of the sex-chromosomes the female is XX, the male is XY.

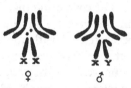

Fig. 1. Conventional diagram of the chromosomes of *Drosophila melanogaster*. (*After* Morgan.)

The situation thus arises that in all the cells of the body of the female and in her immature ova there are four pairs of homologous chromosomes, and of these one pair consists of two X-chromosomes, whereas in all the somatic cells and immature gametes of the male there are four pairs and of these one consists of an X-chromosome in association with a Y-chromosome. Into each ripe gamete there passes one or other member of each pair. All eggs are alike in that each contains an X-chromosome but there will be two kinds of sperm, one containing three autosomes and one X-chromosome, the other three autosomes and a Y-chromosome. The female of Drosophila is monogametic; the male is digametic. When egg and sperm unite in fertilisation, into the zygotes will be received one member of each pair from the father by way of the sperm, and the other member of each pair from the mother by way of the egg, and there will be two forms of zygotes, one that received an X-chromosome by way of the sperm and the other that received a Y-chromosome. The first will have a sex-chromosome constitution that can be symbolised as XX, the chromosome constitution typical of a female, the other a sex-chromosome constitution symbolised as XY, that of a male. This sex-determining chromosome mechanism yields results that are in every way comparable with those that are obtained when a heterozygous dominant (Aa) is mated to a recessive (aa) in a typical Mendelian monohybrid experiment—equal numbers of the two classes that were represented in the mating. In respect of the X-borne genes, the male is constitutionally simplex, the female duplex.

The conclusion that a difference in the gametes of the two sexes is correlated with the sex of the future individual is abundantly supported by the results of other

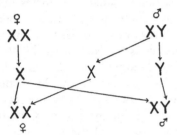

Fig. 2. The XY sex-determining mechanism.

cytological research. As early as 1902, McClung was able to demonstrate that in various Orthoptera (the crickets, cockroaches, and grasshoppers) there is in the male an unpaired chromosome instead of the equal pair in the case of the female. The constitution of the female in these forms is XX, of the male XO, so that while all the female gametes must contain an X element, only half of the spermatozoa will do so. An egg fertilised by an X-bearing sperm must give rise to an XX type of individual, a female. An egg fertilised by a sperm which lacks the X will give rise to an XO type of individual, a male. In other forms it was found by Wilson (1906) and others that the number of chromosomes was identical in both sexes, but that, as in the case of Drosophila, while in one sex the X was equally paired, in the other it was paired with an unequal mate; the female is XX, the male XY in some groups; in others, in *Abraxas* and birds for example, the female is XY and the male XX.

In a third type of sex-chromosome constitution the X is represented, not by one chromosome, but by a group of any number from 2 to 8, which during gametogenesis act together as a compound X-element. The group is single in the male, double in the female, and these cases, therefore, conform to the XX = ♀, X = ♂ type. If, as in certain cases there is a Y-chromosome in addition, this is always single. In other forms the X-chromosome is found to be joined to the end of one of the autosomes, remaining constantly associated with the autosome throughout the whole of the chromosome cycle.

(For comprehensive lists showing the sex-chromosome constitution of different species see Harvey 1916, 1920 and Wilson 1925.)

Sperm dimorphism, estimated by measuring the head length of the sperm, has been demonstrated in a considerable number of species (see Wilson 1925). It is found that there are two intergrading size classes, and it is assumed that the larger consists of the X-bearing, the smaller of the no-X-bearing sperm.

In the case of those forms in which hermaphroditism is customary no satisfactory body of evidence concerning the cytological concomitants of this condition is yet available. In such forms as the oligochaetes, leeches, pulmonates, and ascidians, actual sex-chromosomes have not yet been identified, so that it has been impossible to demonstrate any correlation between chromosomes and sexuality. One instructive and significant case among the normally hermaphroditic nematodes exists, however. Boveri (1911) and Schleip (1911) have studied *Angiostomum* (*Rhabditis*) *nigrovenosum*, a nematode exhibiting an alternation of generations between a free-living dioecious form and a parasitic hermaphroditic form which lives in the lungs of the frog. In the dioecious generation males and females occur in approximately equal numbers and the zygotes produced by these develop into a hermaphroditic form which has the general structure of a female with a gonad that first elaborates oogonia and then later spermatogonia also, the two kinds being produced in irregularly alternating zones. From the spermatogonia spermatocytes arise and these undergo the two usual divisions to form functional sperm. The eggs are fertilised by the sperm elaborated by the same individual with the result that approximately equal numbers of males and females are produced.

Cytological examination showed that the parasitic hermaphroditic form is diploid containing twelve chromosomes, and that this number is reduced to six in the egg, whereas of the sperm only one half receive six, the other half getting only five, owing to the fact that in the second division one of the X-chromosomes remains near the equator and, failing to enter the daughter nucleus, degenerates. Fertilisation results, therefore, in two kinds of zygotes, one possessing twelve chromosomes, the other eleven. Males of the dioecious generation have eleven chromosomes, so that of the sperm elaborated by these one half carries six, the other only five. All the fertilised eggs, however, produce hermaphrodites with twelve chromosomes. To explain this, Boveri and Schleip suggest that the five-chromosome-containing sperm is non-functional, but, on the other hand, it may be that the eleven-chromosome-containing zygotes are non-viable.

THE EVIDENCE OF PARTHENOGENESIS.

Parthenogenesis occurs naturally in many species of plants and animals, *e.g.* the rotifers, daphnids, and such insects as the aphids, gall-flies, and bees, and in certain instances can be induced artificially by physico-chemical means in the sexual ova of forms in which parthenogenesis is not customary. In the rotifers, daphnids, ostracods, aphids, and phylloxerans, the parthenogenetic ovum is an agamete incapable of becoming fertilised, and in its mode of maturation and in other ways is markedly different from the sexual ovum although it is commonly held that the agamete is derived from the gametic form of ovum.

Natural parthenogenesis occurs in two forms, (1) that in which the chromosome number in the parthenogenetic egg is diploid, and (2) that in which the number is haploid, though in a few cases the distinction is bridged in that development begins with the haploid number of chromosomes, but ends with the diploid. The two types are to be distinguished by the number of polar bodies formed during the phase of maturation, the diploid egg extruding but one polar body, the haploid producing two. The haploid egg is one that undergoes complete reduction and is capable of fertilisation, but is one that may develop without activation by the sperm. Such an egg, developing parthenogenetically, yields an individual with the haploid number of chromosomes ($1N$), and so far as is known such an individual is always a male. The diploid egg undergoes no reduction of chromosome number so that the egg possesses a chromosome number equal to that of the ordinary sexually produced zygote, the diploid ($2N$). The diploid eggs are, as far as is known, incapable of being activated by the sperm. If such a diploid egg were so fertilised, a $3N$ zygote would result.

Artificial parthenogenesis has been induced in the eggs of various animals, such as sea-urchins, starfish, annelids, molluscs, and frogs, through the employment of different procedures. The case of the frog possesses special interest in that traumatic partheno-genesis, induced by puncturing the egg periphery with a fine needle, yielded tadpoles that passed through the metamorphosis and became sexually mature. These included 15 males, 3 females, and 2 doubtful (Loeb, 1918).

In the aphids the fertilised egg yields a female which is partheno-genetic, being the first of a series of similar females all of which reproduce by diploid parthenogenesis. Later in the summer time parthenogenetic stem mothers make their appearance, and their offspring, all parthenogenetically reproducing, are of two types, one yielding only sexual females, the other only males. The gametes elaborated by these in their turn initiate a new cycle producing in the following spring a new line of parthenogenetically producing females. Such a life history characterises also the phylloxerans, the daphnids, and the ostracods, and in all (with one exception) the parthenogenesis is of the diploid type. In the aphids and in some phylloxerans there is an "open life cycle," there being an indefinite number of parthenogenetic generations before the sexual forms appear. In other species of the phylloxerans there is a "closed life cycle," the stem mother arising directly from the zygote and yielding two kinds of offspring, one producing only large "female-yielding" eggs, the other only small "male-yielding" eggs, and, as is usual, the cycle ends with fertilisation.

In the case of the bee, wasp, and ant, parthenogenesis takes the haploid form and no alternation occurs, all fertilised eggs yielding females, those not fertilised giving rise to males. The females are diploid, the male haploid in chromosome constitution.

In the rotifers both kinds of parthenogenesis are found. The early part of the life cycle is as that of the aphid; a series of diploid female-producing and parthenogenetically produced generations is followed by a generation, the eggs of which may develop after fertilisation or by haploid parthenogenesis. From this stage, the history is as that of the bee, for if the egg is fertilised, it yields a diploid female, if it is not, it yields a haploid male.

In *Neuroterus*, the gall-fly, all fertilised eggs yield females, but some of these by diploid parthenogenesis give rise to sexual females, whereas others yield males by haploid parthenogenesis, as is the case in the rotifers and in the hymenoptera.

In the case of the forms in which the female is monogametic, it is seen that the fertilised egg always yields a diploid female; that in diploid parthenogenesis the offspring is either a male or a female; and that haploid parthenogenesis yields only males. The fact that diploid parthenogenesis yields both males and females is to be explained in certain cases, such as the aphids and phylloxerans (Morgan, 1908, 1909; Baehr, 1908, 1909), by the fact that the male has one or two chromosomes less than the female. The production of the male, therefore, involves the elimination of one or two chromosomes, it demands that reduction shall occur in the case of one or two pairs of homologous pairs of chromosomes. Morgan (1909) was able to show that in the case of two species of phylloxerans in which the X-chromosome consists of two usually separated elements; during the maturation of the smaller male-producing parthenogenetic egg two chromosomes fail to enter the egg nucleus and are probably extruded into the polar body. It is reasonable to assume that the two extruded chromosomes represent two elements of an X-chromosome, the mate of which remains in the egg, so that whilst the parthenogenetic egg is diploid in respect of the autosomes, the one, the female-producing, will have the sex-chromosome constitution XX, the other, the male-producing, one X; a condition of things exactly similar to that which obtains in the case of sexual reproduction in forms in which the female is XX, the male XO or XY.

In certain forms in which parthenogenesis is quite common, sexual reproduction also occurs, and in these it is found that should an egg be fertilised, it always yields a female, thus differing from ordinary sexual reproduction when fertilisation yields both males and females. The reason for this is that only the X-chromosome-bearing sperm, *i.e.* the female-producing, proceed to complete development. In the first spermatocyte division the X-chromosome passes to one pole but the division is markedly unequal so that two secondary spermatocytes are formed, one being larger and containing the X, the other smaller and containing no X. The latter fails to divide further and degenerates, and the only sperm that are elaborated are those that contain an X (phylloxerans, Morgan, and aphids, v. Baehr, 1908, 1909).

In the case of the phylloxerans the egg acquires the appropriate chromosome number during maturation, the larger ones retaining the full diploid number, the smaller eliminating one X. The eggs are predetermined as male- or as female-producing irrespective of their chromosome number, but they do not become so determined until maturation has led to the establishment of the appropriate chromosome content.

In the case of the parthenogenetically produced frogs reared by Loeb, Parmenter (1919) examined many of these and found that all of them as well as a normal male possessed the diploid number of chromosomes, about twenty-six.

The classical observations of Dzierzon (1845) concerning sex-determination in the honey-bee have been amply confirmed by more recent work, and it is known that the condition in which females alone are produced from fertilised eggs, and males alone from unfertilised, is widespread among the Hymenoptera. The eggs of bees will develop without fertilisation, but should they do so, the resulting individuals are invariably males. Fertilised eggs, on the other hand, usually become females (queens and workers). The sex of the individual depends on whether fertilisation occurs or not, and thus is established before the inception of embryonic existence. The cytological aspects of this condition, however, are even yet somewhat confused. It seems to be certain that the female possesses the diploid number of chromosomes or some multiple of this, whilst the male possesses the haploid. Though the sex-chromosomes have not yet been identified as such, all the evidence that exists points to the conclusion that the female is XX, the male X, and that this number is not doubled during the course of development (Wilson, 1925).

In *Hydatina senta* from the fertilised egg a female arises and she produces parthenogenetically offspring which are all females. Of these some produce only male offspring, others only females. If a female is impregnated within a few hours after hatching she may produce sexual or winter eggs clearly distinguishable from the parthenogenetic eggs (Maupas, 1890; Shull, 1910). It was noted that where young females were given every opportunity to copulate no males were produced and that about the same percentage of females produced sexual eggs as produced males when copulation was prevented. From this it was concluded that the winter eggs were fertilised "male" eggs and that young females destined to produce females cannot be fertilised. In *Hydatina* the sexual female is not to be distinguished from the parthenogenetically reproducing female yet the two forms differ fundamentally, for only the form which, reproducing parthenogenetically, elaborates eggs that are destined to become males is able to elaborate the sexual egg which is a fertilised "male" egg. The sexual female always appears one generation earlier than the male for she is, if unfertilised, the mother of the males. Fertilisation therefore actually changes the destiny of the egg, for in the absence of fertilisation such an egg becomes a male, following fertilisation it develops into a female.

In an American parthenogenetic strain of the rotifer *Hydatina senta*, it has been shown (Whitney, 1914, 1915, 1916) that a continuous diet of the colourless protozoan flagellate *Polytoma* causes practically all females to be produced, and that if the diet is suddenly changed to an active green protozoan flagellate *Chlamydomonas*, males can be produced in great numbers.

Shull (1910, 1911, 1912, 1913, 1915); Punnett (1906); Maupas (1891); Mitchell (1913); and Nussbaum (1897) have also studied the relation of nutrition to sexuality in rotifers. Shull has demonstrated in *Hydatina* that the sex of an individual is determined a generation in advance and that it is not so much the amount of food available but the amount assimilated during the period when reproduction and growth are rapid that seems to be the deciding factor. He argues that it is probable that both high nutrition and male production are the results of some other as yet unknown physiological factor, and that it is incorrect to assume that male production is the result of nutritional change.

The Cladocera elaborate two kinds of eggs, those that require to be fertilised and those that without fertilisation give rise to males. Banta and Brown (1923, 1924) agree with others that it is reasonable to assume that the sexual form is called forth by environmental agencies. The simple expedient of overcrowding the females, as· was previously noted by Smith working with *Moina rectrirostris*, led to the production of males. It was found that this was not due to scarcity of food, the age of the medium, or to the accumulation of excretory products, but that possibly an accumulation of CO_2 or a depletion of oxygen was the responsible agent.

Riddle (1916) has suggested that changes in metabolic rate are at the bottom of the results which have been obtained in the case of the rotifers. In the case of the Cladocera, Banta and Brown are of opinion that it is not a matter of a change in metabolic rate but rather that a lower rate is associated with male determination in the parthenogenetic eggs and a higher rate with female determination.

Schrader (1925) described a form of Daphnia in which the sexual, or winter, egg develops without fertilisation. No males have as yet been found. There is apparently no synaptic stage in the prophase. There is only one maturation division and no reduction in the number of chromosomes. The parthenogenetic, or summer, egg, although morphologically quite distinct, has a similar process of maturation and the same chromosome number. Storch (1924) and Whitney (1924) recorded similar observations in the rotifer *Asplanchna*. Fertilised sexual eggs with the diploid number of chromosomes develop into females; unfertilised, with the haploid number, into males; while the mature parthenogenetic egg retains the diploid number and develops into a female.

The genetic and cytological evidence concerning parthenogenesis points to the conclusion that sex-determination is an affair of the chromosomes, particularly of the sex-chromosomes, for it is seen that there is a most significant relation between differences in chromosome numbers in zygotes and differences in sexuality. The mechanism of sex-determination in parthenogenetically reproducing forms is identical with that which operates in normally sexually reproducing forms, though the mode of its functioning may be different.

GENETIC EVIDENCE.

A. SEX-LINKED INHERITANCE.

(1) *In* Drosophila melanogaster.

In the case of *Drosophila*, the son gets his single X from his mother, the daughter gets one from the mother and another from the father. If on the X-chromosome of the father is borne the gene for a recessive character, the track of this chromosome in inheritance can be followed if the Y-chromosome of a male carries either no genes at all or else none that affect the action of the X-borne genes, and since it can be shown that the Y can be absent without the processes of sex-determination being affected, as is seen in the case of *Metapodius* described by Wilson (1909), this assumption is not extravagant. As an example of sex-linked inheritance, the case of mating red-eyed and white-eyed flies may be taken. The eyes of the wild fly are red; in one culture a mutant white-eyed fly appeared. He was mated to a red-eyed female, and the offspring, both males and females, were red-eyed. These were interbred and produced in the next generation on the average three red-eyed to one white-eyed in every four, equal numbers of males and females, *but every white-eyed individual was a male*. The white-eyed grandfather had transmitted the character white-eye to half his grandsons but to none of his granddaughters. A white-eyed female can be produced easily by mating the F_1 heterozygous red-eyed female to a white-eyed male. When a white-eyed female was mated to a red-eyed male, the reciprocal mating that is, *all the daughters were red-eyed and all the sons were white-eyed*; when these were interbred, they gave red-eyed males and females, and white-eyed males and females in equal numbers. These results can be explained easily and satisfactorily if it is assumed that the genes for white-eye and for red-eye are X-borne, that the characters white and red are allelomorphic, and that the Y-chromosome of the male bears no genes at all.

In the following diagrams the X-chromosome is represented by a stout rod with rounded ends. The solid X-chromosome is that which bears the gene for the character red-eye.

Many other characters of *Drosophila* are inherited according to the above scheme. These sex-linked characters (not sex-limited since they can be exhibited by both sexes) constitute the first linkage group, and their genes are likewise placed upon the X-chromosome.

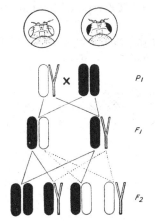
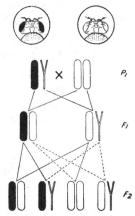

Fig. 3. White-eyed ♂ × Red-eyed ♀.
(*After* Morgan.)

Fig. 4. Red-eyed ♂ × White-eyed ♀.
Criss-cross inheritance in F_1.
(*After* Morgan.)

(2) *In the Cat.*

This type of inheritance was first recognised in the cases of the human subject and of the cat. It is commonly held that the tortoise-shell coat colour in the cat is a typical sex-linked character.

It is accepted that the mating black ♂ × yellow ♀ gives tortoise-shell ♀♀, yellow ♂♂; that yellow ♂ × black ♀ gives tortoiseshell ♀♀, black ♂♂, and an occasional black ♀; that black ♂ × tortoiseshell gives tortoiseshell ♀♀, black ♀♀, and yellow ♂♂, and black ♂♂; and that yellow ♂ × tortoiseshell ♀ gives yellow ♀♀, tortoiseshell ♀♀, yellow ♂♂, black ♂♂, and an occasional black ♀.

If this is so, it follows, as Little (1920) has pointed out, that black and yellow in the cat are sex-linked characters and that together they constitute the tortoiseshell character. The facts can be explained if the following assumptions are made. The allelomorphic factors B (black) and b (yellow) are resident in the X-chromosomes, and the Y-chromosome of the male does not carry these factors. The female, possessing two X-chromosomes, can have one bearing the factor for black, another bearing the factor for yellow; she can be tortoiseshell. The male can have either the factor for black or for yellow in his single X-chromosome. He cannot be tortoiseshell. It will be noted that in this interpretation it is assumed that in the cat the male has the XY type of sex-chromosome constitution, not the XO.

This scheme does not account for the exceptional black female that appears among the offspring of matings 2 and 4, nor for the

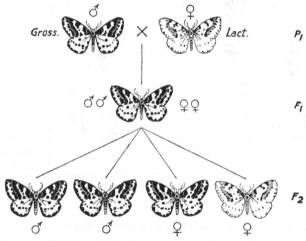

Fig. 5. *Grossulariata* ♂ × *lacticolor* ♀. (*After* Doncaster.)

fact that a sterile tortoiseshell male very exceptionally makes its appearance. The occasional black female is possibly due to the occasional complete expression of black in the simplex state, and the tortoiseshell male to crossing-over between the X and the Y in a yellow male and between the X's in females.

(3) *In* Abraxas.

Sex-linked inheritance was first studied by Doncaster (1908) in the currant moth, *Abraxas*, two varieties of which, var. *grossulariata* and var. *lacticolor*, are distinguished by the colour pattern of the wings. When a *lacticolor* female was mated to a *grossulariata* male, all the F_1 individuals were *grossulariata*; the *grossulariata* pattern is therefore dominant to the *lacticolor*. In the F_2 both types were found in the proportion of 3 *grossulariata* to 1 *lacticolor*, but *all the lacticolors were females*. When the F_1 heterozygous *grossulariata* male was back-crossed with *lacticolor* females, the result was the expected one, *e.g.* equal numbers of both sexes of *grossulariata* and *lacticolor* forms. When one of these *lacticolor* males was mated to an F_1 *grossulariata* female, equal numbers of *grossulariata* and *lacticolor* forms resulted, but all the former were males and all the latter females.

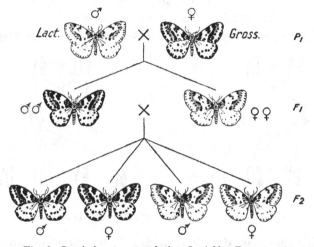

Fig. 6. Lacticolor ♂ ×grossulariata ♀. (*After* Doncaster.)

When a *lacticolor* male was mated to a *grossulariata* female, the males in F_1 were all *grossulariata*, the females all *lacticolor*. Reciprocal crosses thus gave different results.

This can only be interpreted on the assumption that homozygous *grossulariata* males transmit the *grossulariata* character to the off-spring of both sexes, while a *grossulariata* female transmits it only to her sons. Hence it would seem that the female produces two sorts of gametes, those which bear the factor for the *grossulariata* character and are destined, if fertilised, to become males; and those which do not carry this factor and are destined to become females. The female is then constitutionally heterozygous for the character *grossulariata*, and the factor for this character is most intimately associated with another which, if present in the duplex state, determines maleness. The results of further breeding completely confirm this hypothesis.

These results can be shown diagrammatically as follows, assuming that the female in *Abraxas* has a single X-chromosome, the male two, and that the genes for the sex-linked characters are resident in the X, as appeared to be the case in *Drosophila*.

The son receives one X-chromosome from his father and one from his mother. In the reciprocal cross (Fig. 8) as the latter X carries the factor for *grossulariata*, he will exhibit the *grossulariata* character. The daughter has but a single X-chromosome and this she receives from her *lacticolor* father, who must be homozygous

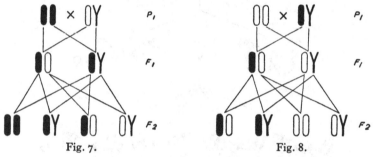

Fig. 7.   Fig. 8.

Chromosome interpretation of sex-linkage in *Abraxas*.

for this character, since *grossulariata* is dominant and would be exhibited even if the factor for this character were present only in the simplex state. All the daughters therefore will be *lacticolor*.

From the study of the transmission of these sex-linked characters the conclusion emerges that the sex of *Abraxas* is decided by the simplex or duplex condition of some component, which, when present in duplicate, leads to the establishment of maleness. The evidence suggests that there is a sex-determining mechanism, an XY-mechanism, that the sex-chromosome constitution of the male is XX, that of the female XY, and that the factors for the sex-linked characters are borne upon the X-chromosome as are also those which when present in duplicate lead to the production of males.

Doncaster examined *Abraxas* cytologically and found that both male and female had 56 chromosomes and that in the female no obviously unequal pair could be distinguished. It is thus seen that the evidence for an XY mechanism may be derived solely from breeding experiments involving sex-linked characters, and that the X- and the Y-chromosomes may not be different in shape and size. However, Doncaster (1914) found that in one line of his *Abraxas* many of the females gave none but female offspring. Theoretically this would be expected if these females possessed no X-chromosome. On examination it was found that these females had 55 instead of 56 chromosomes and so during the maturation divisions of the ova 28 chromosomes foregathered at one pole, 27 at the other, so that the mature ovum contained either 27 or 28 (27 + the Y). The sex-chromosome content of these eggs is either O or Y, and these are fertilised by an X-bearing sperm to yield XO and XY zygotes, all females. The Y-chromosome in *Abraxas* is as functionless as it is in the case of *Drosophila* as far as sex-determination is concerned. By tracing the distribution of visible characters, it is thus possible to follow the transmission from generation to generation of "sex-determining" factors, the fate of which is linked with that of those corresponding to the *grossulariata* and *lacticolor* characters. In *Abraxas* the female is the digametic sex, elaborating, as far as the elements of the sex-determining mechanism are concerned, two sorts of gametes; the male is monogametic.

(4) *In the Fowl.*

A precisely similar mode of inheritance of sex-linked characters is found in birds, and this fact, enunciated so ably by Punnett, has been taken advantage of by poultry keepers, since it enables them to tell the sex of chickens at the time of hatching.

Barred (dominant) and non-barred (recessive) for example are the members of a sex-linked pair of characters. A black (non-barred) cock (bX) (bX) mated with barred hens (BX)Y will throw barred sons and black (non-barred) daughters, and since the chicken which is to become a barred adult is black but with a characteristic whitish spot on the top of the head in addition to the white underparts, whereas the day-old chick that is to become a black (non-barred) bird has not the white head-spot, it is possible to distinguish the sex of any bird in the F_1 of this mating at the time of hatching. The size and distinctness of the head-spot are very variable; it may involve but a few down feathers, or it may be quite large, but this is of no importance. If there is a yellowish-grey spot on the top of the head of a chicken in this mating, that chicken is a male and will be barred.

B. NON-DISJUNCTION.

Thorough investigation of the phenomenon of *non-disjunction* finally placed the correlation between sex-determination and the distribution of the X- and Y-chromosomes on an unshakable foundation. The work of Bridges (1916), in a most spectacular way demonstrated the precision with which the distribution of the chromosomes and of sex-linked characters coincides. In his non-disjunction work Bridges actually dealt with the mutant eye-colour character "vermilion," but "white" eye-colour may be used as an example. It will be remembered that the mating red-eyed ♂ and white-eyed ♀ of *Drosophila* produces white-eyed ♂♂ and red-eyed ♀♀, so

$$P_1 \quad (WX) Y \quad \times \quad (wX) (wX)$$
$$F_1 \quad (WX) (wX) \quad : \quad (wX)Y$$

During the course of certain experiments there appeared in F_1, in addition to the usual two classes of offspring, certain exceptional white-eyed females and red-eyed males; when such a white-eyed female was mated to a red-eyed male, again all four classes appeared in the resulting FD_2, *i.e.*, white-eyed females, which when bred behaved in a similar abnormal fashion; red-eyed females, of which half gave normal and half exceptional results in further breeding; red-eyed males behaving normally; and white-eyed males, of which some produced the usual sorts of offspring and some produced exceptional daughters. On cytological examination of such exceptional white-eyed females, Bridges found that the dividing nuclei of their cells displayed a Y element in addition to the normal pair of X's. The condition was that of *secondary non-disjunction* and can be interpreted on the assumption that at reduction of the egg in which this individual had its origin, the XX pair in exceptional cases failed to *disjoin* so that the mature ovum contained either two XX or none. An egg which possesses two

X's instead of one may be fertilised by an X-bearing or a Y-bearing
sperm and so the resulting zygotes may come to have three X
elements or two X's and a Y. An XXX individual is a female; an
XXY individual is also a female although it possesses a Y-chromosome in addition to its two X's. An egg which possesses no X-
element at all can be fertilised by an X-bearing or a Y-bearing
sperm to form an XO or an OY individual, and of these an XO
individual is a male, quite normal in appearance but sterile; an
OY does not develop. The type of non-disjunction consequent
upon the failure of the two X-chromosomes to disjoin is known
as *primary non-disjunction* of chromosome I. The type of non-
disjunction consequent upon the presence of an extra sex-chromosome is known as secondary non-disjunction of the sex-chromosomes. Except for an unusual sex-ratio the presence of such
(cytologically) exceptional males and females would not be suspected
in one and the same strain; but should such a non-disjunctional
female be employed in a sex-linked experiment the appearance
of exceptional phenotypes would indicate what had happened.
For example, the mating of a red-eyed male and a white-eyed
primary non-disjunctional female would give the following results:

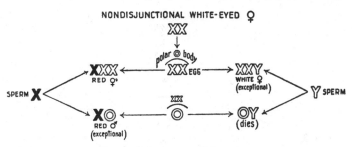

Fig. 9. Primary non-disjunction.

In the case of the XXY white-eyed female encountered by Bridges
it is to be assumed that when homologous chromosomes pair in
synapsis, two types of reduction division are possible. If the X's
conjugate, then in reduction they disjoin and pass to opposite poles
and the Y-chromosome will pass to one or the other pole. Thus X
and XY ova will be produced in equal numbers. If, on the other
hand, an X conjugates with the Y, then X and Y pass to opposite

poles where one of them will be joined by the other X. Thus, X, XX, XY, and Y ova will be produced. From experimental evidence it has been determined that in non-disjunctional females, *homo-synapsis* (the pairing of the two X-chromosomes) occurs in 84 per cent. of cases and *hetero-synapsis* (the pairing of an X with the Y) in 16 per cent.:

			XXY				
Homo-synapsis 84 %	XX	Y	:	XY		X	*Hetero-synapsis* 16 %
Reduction	X 42 %	XY 42 %	:	X 4 %	Y 4 %	XX 4 %	XY 4 %	Reduction
Ova	(X)^{xy} 46 %	(XY)^{x} 46 %	:	(XX)^{y} 4 %		(Y)^{xx} 4 %	Ova

The ova of an XXY white-eyed non-disjunctional female may then be of four sorts associated with four sorts of polar bodies. If this female is white-eyed there can be no gene for the red-eyed condition in her genetic constitution and the genes for white-eye are borne upon the X-chromosomes. If her ova are fertilised by the sperm of a red-eyed male (the gene for red-eye being carried on the single X-chromosome) the history of her chromosomes which bear the gene for white-eye can be followed.

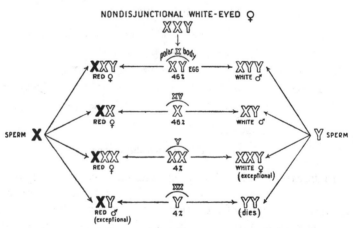

Fig. 10. Secondary non-disjunction. (*After* Bridges.)

The exceptional white-eyed daughters of the non-disjunctional XXY female and the red-eyed XY male are white-eyed because they do not get one of their X-chromosomes from their father; the exceptional red-eyed males are red-eyed because they get their Y-chromosome from their mother, and their X from their father. Thus, the non-disjunction of the X-chromosomes can explain the entire series of the exceptional genetic phenomena which occur in these strains. When once a non-disjunctional female is present in a stock, unusual results must accumulate in increasing proportions. The experimental breeding results, endorsed by the cytological evidence, turned what seemed to be in direct contradiction to the interpretation of sex-linked inheritance in terms of the chromosome theory of heredity into a most spectacular confirmation of this theory. A consideration of the results of non-disjunction will show how chromosome aberration can lead to dissimilarity in the characterisation of closely related individuals. The irregularities in the distribution of the chromosomes may be of various kinds and occur in all probability during the maturation division of the gametes. The migration of the chromosomes to the poles of the dividing cell may be irregular, so that the two daughter-cells come to possess an abnormal number. The essential feature in chromosome aberration is the quantitative abnormality in the chromosome content as opposed to the qualitative abnormality in the case of mutation.

The Y-chromosome in *Drosophila*, a partner to the X-chromosome during gametogenesis, is not concerned in the determination of sex. It would seem that one X, or one "dose" of some sex-determining gene or genes upon the X, normally results in the production of a male, two "doses" producing a female. Sex-linked characters are associated with the sex-determining mechanism because their genes are located in the sex-chromosomes and these characters do not necessarily have anything to do with the sexual organisation of the individual; they are sex-linked, not sex-limited characters, and are mainly concerned in the general development of the body as are most of the characters the genes for which are placed upon the other chromosomes. The facts of sex-linked inheritance point to the conclusion that sex is determined at the time of fertilisation and that the XX-XY-chromosome mechanism is the sex-determining mechanism, providing in each generation equal numbers of males and females.

Chapter II

THE MECHANISM OF SEX-DETERMINATION
(*continued*)

GENETIC EVIDENCE (*contd.*)

C. GYNANDROMORPHISM.

The clearest light is thrown upon the sex-chromosome sex-determining mechanism by the phenomenon of gynandromorphism, which is an intersexual condition due to a regional disharmony in the distribution of the sex-chromosomes.

(1) *In* Drosophila melanogaster.

A gynandromorph is an individual of a bisexual species which exhibits a mosaic of male and female sexual characters; it is a sex mosaic in space. In the case of *Drosophila melanogaster*, about 1 in every 2000 individuals exhibits this condition of gynandromorphism. Most of these are bilateral gynandromorphs exhibiting the complete male characterisation on one side of the antero-posterior mid-line of the body, the complete female characterisation on the other, with a sharply demarcated line of junction of the two kinds of tissue. Since the male body is normally smaller than that of the female the gynandromorph's body is bent towards the male half. Usually in these cases there is an ovary on the female side, a testis on the male, but this is not always the case, nor would it be expected that it should be, for as Huettner (1922) has shown, the gonads are not formed from a single nucleus but from several nuclei which give rise to the primordial germ cells, and so it is not inevitable that both gonads should be histologically and cytologically similar. Commonly, however, there are two ovaries or two testes in a bilateral gynandromorph. In other cases of gynandromorphism one-quarter of the body is male in its sexual characterisation, three-quarters female; in still others less than a quarter is male while more rarely the head is female whilst the rest of the body and abdomen are male.

Morgan and Bridges (1919) have described many of these

sexually abnormal forms in great detail and have shown that if in the mating that produces the gynandromorphic forms sex-linked characters are involved, and if the sex-linked characterisations of the two parents are dissimilar, then the sex-linked characters of the male parts are those exhibited by the father or those exhibited by the mother, whereas the sex-linked characters of the female parts are a combination of the sex-linked characters of both parents, and that, in respect of the autosomal (non-sex-linked) characters, male and female parts are alike. The example depicted below is a bilateral gynandromorph, the left side exhibiting a typical female

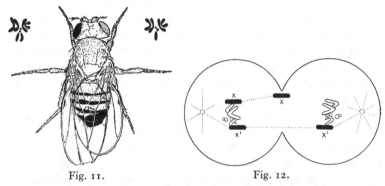

Fig. 11.   Fig. 12.

Fig. 11. A bilateral gynandromorph. (*After* Morgan, Bridges and Sturtevant.)
Fig. 12. Diagram illustrating Morgan's elimination interpretation of gynandro-
          morphism. (*After* Morgan, Bridges and Sturtevant.)

characterisation, the right a typical male. The left side presents the dominant sex-linked character, Notch (the mother was Notch), and the right exhibits the recessive sex-linked characters scute, broadwing, echinus eye, ruby eye, tan body-colour, and forked bristles (the father had these characters).

These facts point to the conclusion that gynandromorphism in such a case as this results from aberration in the distribution of the X-chromosomes. If it is assumed that the gynandromorph is in its beginning an XX-zygote, a genotypic female, and that at some stage during the early cleavage divisions of the fertilised egg a daughter X-chromosome fails to enter one of the daughter cells, this cell will then contain one X instead of two, i.e. will contain the genotype of a cell that normally is XO or XY. Since one X is derived from the paternal contribution of nuclear material, and the other from the maternal, and since, if the parents were dissimilar in their sex-linked characterisation these two X-chromosomes will be dissimilar in their gene content, it follows that the characterisation based upon genes in either of them can be markedly different from that based upon the genes in both.

The mother (XX) carries in each of her X-chromosomes the gene for the dominant sex-linked mutant character Notch. The father (XY) carries in his single X-chromosome (X') the genes for the recessive sex-linked characters scute, broad, echinus, ruby, tan, and forked. The F_1 female zygote will receive one X-chromosome from each parent. If nothing untoward happens during ontogeny she will exhibit the dominant character Notch and not scute, not broad, not echinus, not ruby, not tan, and not forked—the dominants of both parents—and be a typical female. If, however, during the first cleavage division of this zygote the maternal X-chromosome fails to enter a daughter cell, the paternal X' alone will be present therein and the tissues that develop from this daughter cell will be $1X : 2A^1$ (male) in constitution and cannot be Notch but will exhibit, if they belong to the appropriate parts, the recessive sex-linked characters of the father. If it is the paternal X' that is thus eliminated, the male side of the gynandromorph will exhibit the sex-linked character of the mother, Notch. In the example cited it is seen that it was the maternal X that was eliminated.

In the absence of further aberrations in chromosome distribution, all the cells that have their origin in the XX daughter cell will have the $1X : 1A$ type of chromosome relationship; all the cells that arise from the single X-containing daughter cell·will have the $1X : 2A$ type, and the sex-linked characters of the tissues formed by these latter, if they pertain to parts that do exhibit sex-linked characters, will be those of the parent from whom their X-chromosome was derived. If this is the case, then a gynandromorph of a grade that is compatible with fertility, i.e. one in which the abdomen and genitalia are female, should prove to be heterozygous in respect of her sex-linked characterisation. This is the case. On the other hand, if the abdomen and genitalia are of the male type in their architecture, the individual should be sterile, since in these parts

---

[1] Where $1A = 1$ complete set (haploid) of autosomes.

the sex-chromosome constitution is XO, and it is known from Bridge's non-disjunction work (1916, 1919) that the XO male in *Drosophila* is sterile. This is also the case.

If the elimination of the X-chromosome occurs at the first cleavage division of the zygote, the gynandromorphism is lateral; if it occurs at the second, one-quarter of the body will be male in its characterisation. Elimination may occur at any stage of ontogeny and the later it occurs the smaller will be the area that exhibits male type characterisation. But however large or small this area may be, it will pursue its development under the direction of its own genotypic constitution (except in the case of vermilion eye, Sturtevant, 1920) and the course of its differentiation is not influenced by the physiological activity of the gonads.

It is to be noted that these mosaic formations are not limited to the sexual phenotype: they are commonly found involving the general phenotype in hybrids. Gynandromorphism results only from such aberrations in the distribution of nuclear chromatin that lead to a sufficient disturbance in the X : A ratio.

(2) *In* D. simulans.

Gynandromorphs have occurred fairly commonly in *Drosophila simulans* (Sturtevant, 1921). Most of these are to be interpreted as the result of the elimination of an X-chromosome at a cleavage division, but one specimen was a mosaic both for sexual characters and for mutant characters of chromosome III. Analysis of this case showed that in its production two spermatozoa were involved and that it was the result either of the synchronous fertilisation of a double nucleus egg by two sperms or of polyspermy in a single nucleus egg.

(3) *In the Bee.*

There are certain facts relating to sex-determination in the bee that are unique. The queen is diploid and this double set of chromosomes is reduced to the haploid number in the ripe egg after two polar bodies have been extruded. If the egg is fertilised, it gives rise to a female (queen or worker), but if it is not fertilised it yields a male (drone). The male has the haploid number of chromosomes, but whether sex-determination is the result of the action of a sex-chromosome mechanism has not yet been determined. It is necessary to assume that possibly the sex-determining mechanism in the bee and similar forms is essentially different from that which obtains in the case of *Drosophila*. It is remarkable that the male is a haploid individual and that the diploid individual is not a male for the ratios among the sex-determining genes are not different in the two cases. The case of the bee is so unique that it merits further investigation.

In the bee the mother transmits her characters directly to her sons as is the case in the typical sex-linked mode of inheritance in *Drosophila*. But the reason for this is that the male develops from an egg that has received no paternal chromatin. There is certain reliable evidence which strongly suggests that there is a chromosome in the bee equivalent to the X-chromosome of *Drosophila*. Hybrid queens produce two kinds of males and two only, which fact indicates that a single factorial difference exists and that the allelomorphs are carried by a pair of chromosomes, not necessarily sex-chromosomes, but possibly so. The elimination hypothesis which seeks to explain gynandromorphism in the bee demands that the elimination must involve a sex-chromosome and that this must also be the one that carries the genes concerned in the production of racial characters.

In the sixties of the nineteenth century Eugster kept at Constance a hive of bees which consisted of hybrids out of a yellow Italian queen by darker German drones and which regularly produced a large number of gynandromorphs. This regularity in the production of abnormal forms points to the conclusion that the abnormality is genetic and not merely accidental or due to specific environmental impresses. When the Italian queen died, Eugster replaced her with a hybrid queen and she in her turn produced gynandromorphs. V. Siebold first examined these specimens in 1864, and attempted to account for these gynandromorphs by assuming that an insufficient number of sperms had penetrated the egg so that parts of it lacked sufficient quantities of the male contribution. It is to be noted that Nachtscheim (1913) has recently shown that polyspermy is customary, but this interpretation does not explain all the phenomena encountered. The specimens have recently been re-examined by Boveri (1915) and Mehling (1915). They ranged from an almost perfect male to an almost perfect female.

Since they were hybrids, it was possible to tell from their characterisations whether the tissues concerned contained paternal, maternal, or both paternal and maternal nuclear material. Boveri concluded that all the male parts were purely Italian in their characterisation, whereas the female parts were definitely hybrid, exhibiting the characters of both races. If this were so, it follows that in the tissues of the male parts there was no nuclear material brought into the zygote by the sperm but only that contributed by the ovum.

In order to explain this, Boveri suggested that the sperm did not reach the egg nucleus before the latter had begun to segment and so fused, not with the whole egg nucleus, but with one of the daughter nuclei (= partial fertilisation). Two blastomeres would thus result, one containing only the maternal nucleus whilst the other would include both maternal and paternal nuclear material. It is established that in the case of the bee, if the egg which contains the haploid number of chromosomes is fertilised it becomes a female: if it is not fertilised, it becomes a male. The male is haploid, the female diploid in respect of its chromosome content. It follows then, that in the case of the late fertilised egg, the blastomere that contains both maternal and paternal nuclear material is diploid, *i.e.* female, the blastomere that is not fertilised remains haploid, *i.e.* male.

Morgan has suggested that the same results would follow if two sperms should enter one and the same egg, but only one of them unite with the egg nucleus (= polyspermy). Doncaster (1914) has pointed out that, as seen in his *Abraxas* work, this gynandromorphism could be explained as the result of the synchronous fertilisation of two egg nuclei by two independent sperms.

If this gynandromorphism is the result of partial fertilisation, then the male side should exhibit the characters of the mother's race (Italian) because its haploid chromosomes came directly from the Italian mother egg, whilst the female side should exhibit the dominant characters of both Italian and German races. If, on the other hand, the condition results from polyspermy, then the male parts should exhibit the characters of the father's race (German). Boveri examining the preserved specimens (47 years in alcohol), reached the conclusion that the male parts were Italian in their characterisation. If this is the case, then Boveri's hypothesis is to be preferred to that of Morgan.

Engelhardt (1914), however, has described some fresh gynandromorphic specimens out of an Italian mother by a father very similar indeed to the males of the Eugster hive. The male parts presented the paternal characterisation. This supports Morgan's interpretation.

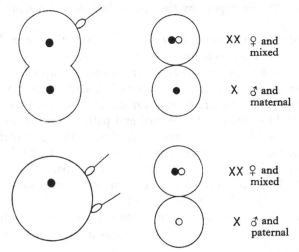

Fig. 13. Boveri's partial fertilisation and Morgan's polyspermy interpretations compared.

Morgan and Bridges are of opinion that the case of Eugster's bees can equally readily be explained by assuming that, as in the case of the *Drosophila* gynandromorphs, there is an accidental elimination of a sex-chromosome during cleavage. The bees concerned were *A. ligustica* (Italian) and *A. mellifica* (German). If it is assumed that one of the postulated sex-chromosomes of the hybrid carried the gene-complex for the *ligustica* characterisation, the other that for the *mellifica*, and if the *mellifica* chromosome is eliminated, then a cell containing only the *ligustica* chromosome would result and all parts descending from that cell would be male (haploid) and *ligustica*. This is the result observed by Boveri. If the *ligustica* chromosome is eliminated the result is male tissue and *mellifica* characterisation. This is the result observed by Engelhardt. This interpretation accommodates both Boveri's and Engelhardt's observations whereas Boveri's partial fertilisation and Morgan's polyspermy each only explains one. If *mellifica* is dominant to *ligustica*, then the female parts will be *mellifica* as described by Boveri. But Newell (1914), who crossed Italian queen and grey Carnolian drone, got yellow daughters, showing that yellow was dominant. In this case both sides would be alike if in the male parts the *mellifica* chromosome had been eliminated, and this may have been the reason for v. Siebold's omission to mention in connection with the fresh Eugster specimens that the parts were different in colour. It is not improbable that when Boveri examined his material the specimens were too old to give reliable pictures of colour differences. Whichever interpretation is correct, and there is no reason to assume that all instances of gynandromorphism result from exactly the same kind of fault, it is certain that in all cases the underlying cause of gynandromorphism is an abnormal distribution of nuclear material; it is the mechanism of distribution that is at fault. According to the stages of embryonic development at which the chromosome distribution is abnormal, the end-result will be bilateral, quartal, or some other grade of sex mosaic.

Nachtsheim (1923) found that in the Phasmidean (Hymenoptera) *Carausius* there occurred about 0·05 per cent. gynandromorphs, predominantly female with more or less demarcated areas of male characterisation. In size these individuals were intermediate between the normal male and female and were less fecund. In this form Nachtsheim found that both sexes were diploid and that though the eggs formed two polar bodies they doubled the haploid number of chromosomes to remain diploid. The progeny therefore all have the same chromosome number and only females appear. The rare male arises through a disharmony in chromosome distribution, possibly non-disjunction, and Nachtsheim argues that the gynandromorph arises also through some chromosome dislocation.

(4) *In the Silkworm.*

Among the Lepidoptera over 1000 cases have been described (see Wenke, 1906). They can all be explained on the assumption that there has been elimination of an X-chromosome in the XX (male) zygote. Toyama (1906) has described two cases in the silk-worm moth (*Bombyx mori*). The mother belonged to a race in which the caterpillar is zebra patterned and the father was of a race in which the caterpillar is uncoloured white. Zebra is dominant to white, and these characters were regarded by Toyama as sex-linked. In the one case the left female side was zebra, the right male was white.

The female is XY in the silkworm. She lays two sorts of eggs, the X-bearing and the Y-bearing. Either of these is fertilised by an X-bearing sperm. Let W = the gene for the zebra pattern, w that for white, and let the brackets represent the fact that the gene is resident upon the X-chromosome included therein.

Boveri's partial fertilisation interpretation.

Egg   Sperm

(WX) × (wX)   This half will develop, will be male (XX) and zebra (Ww).

(WX)
  Y   × (wX)   ,,   ,,   female and zebra.

                              ,,   ,,   female and white.

  Y   This half will not develop.

Morgan's polyspermy interpretation.

(WX) × (wX)   This half will develop, will be male, and zebra.

      (wX)   ,,   ,,   female and white.
  Y   × (wX)   ,,   ,,   female and white.
      (wX)   ,,   ,,   female and white.

Morgan's elimination interpretation.

(WX) (wX)   The zygote.

(WX) (wX)   This daughter cell and the tissues arising therefrom will be male and zebra.

       (WX)   This daughter cell and the tissues arising therefrom will be

or   (wX)   female and zebra if one X is eliminated, white if the other.

(WX)Y   The zygote.

(WX)Y   This daughter cell and the tissues arising therefrom will be female and zebra.

      Y   This daughter cell will not develop.

None of these interpretations accommodates the facts of the case. Fortunately, Doncaster (1914) had described binucleated eggs in *Abraxas*, each nucleus having given off polar bodies and each having been independently fertilised. Morgan, appealing to this fact, points out that if zebra and white are autosomal and not sex-linked characters and if the zebra mother was heterozygous for this character, one nucleus could contain an X-chromosome and an autosome with the gene for white, whilst the other could contain a Y-chromosome and an autosome with the gene for zebra. The two sperms of the father will carry the X and the autosomal gene for white and each fertilising a nucleus would give rise to a male side XX white: white and a female side XY zebra: white. This seems to be the most reasonable interpretation.

Doncaster's binucleated egg interpretation (Morgan).

Ww   XY   The egg.

$$\frac{W \quad Y}{w \quad X} \times wX = \begin{array}{l} Ww\,XY\text{ female : zebra.} \\ ww\,XX\text{ male : white.} \end{array}$$

(5) *In* Abraxas.

The colour patterns *grossulariata* and *lacticolor* are typical sex-linked characters. *Lacticolor* male × *grossulariata* female gives *grossulariata* males and *lacticolor* females. Doncaster (1914) found that one mating produced twenty-four *lacticolor* females, no *grossulariata* males, and one gynandromorph, a *lacticolor* with the right side of the body male and the imperfectly developed left more like that of the female. The internal genitalia were, as far as was known, imperfectly developed male organs.

The absence of males was associated with an unusual chromosome content in this family (55), for all fertilised eggs lacked a chromosome, the single X passing out into the polar bodies in most cases.

Morgan explains this case as follows. The egg either contains no X-chromosome and therefore no gene for *grossulariata* or else it does contain this X and the gene for *grossulariata*. The male in *Abraxas* is XX. A sperm carrying the *lacticolor* gene-bearing X-chromosome fertilising the no-X-egg will yield an XO zygote, a female and *lacticolor*. A similar sperm fertilising an X-bearing egg will yield an XX zygote, a male and *grossulariata*. Only through the fertilisation of a no-X-egg by a two-X-bearing sperm could a *lacticolor* male arise. Such a sperm could result from non-disjunction. If the zygote which developed into this gynandro-morph was such an XX male (both X's having been derived from the father) then the gynandromorphism would be due to the elimination of one of these X's at some early division.

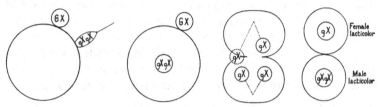

Fig. 14. Morgan's interpretation of Doncaster's case. G=gene for *grossulariata*: g=gene for *lacticolor*.

Doncaster's second case is one that readily lends itself to inter-pretation as an instance of chromosome elimination. *Lacticolor* male × *grossulariata* female gave eleven *grossulariata* males, eleven *lacticolor* females, and one gynandromorph, the anterior parts of which were male and the posterior parts female. The zygote would be an XX individual, one X bearing the gene for *grossulariata*, the other the gene for *lacticolor*. Elimination of a paternal *lacticolor* X-chromosome would give a *grossulariata* male anterior region (XX) and a *grossulariata* female posterior region (maternal X).

(6) *In Birds.*

Since gynandromorphism is the result of an abnormal func-tioning of a mechanism, it cannot be the normal characterisation of a group. It can only affect those tissues the cells of which pursue their differentiation independently of all others. It is, therefore, impossible to find a gynandromorph in those groups in which the control of the processes of sexual differentiation has been com-pletely relegated to some special organ or tissue; it cannot occur in the mammal if in this group sexual differentiation is mainly pursued under the directing stimulus of a sex-hormone elaborated by the differentiated gonad. If in a group undoubted, and other-wise inexplicable, cases of gynandromorphism are found, then in that group the internal secretions of the gonad do not constitute the sole controlling forces in sexual differentiation and the cells of different parts can pursue their differentiation under the direction of their own genotype.

Gynandromorphism has been recorded in birds. Poll (1909) described a true bilateral bullfinch; Weber (1890) a finch (and refers to two other cases described by Cabanis, 1874); and Bond (1913) a pheasant. In the finches the right side of the body was male and the gonad a testis, the left female, and the gonad an ovary. In Bond's pheasant the left side of the body was male, the right female, and the gonads were ovotestes. The individual tail feathers had one half of the vane with male markings, the other half with female. Macklin (1923) describes a fowl (breed not stated) which had the plumage characterisation of a hen with neck-feathering suggestive of that of the male and with tail sickles rather longer than those of the normal hen. (It will be shown later that this tail characterisation is associated with the synchronous presence of both ovarian and testicular tissues in experimental birds, either male or female, and that it is the normal condition in the Campine henny-feathered cock.) The comb and right wattle were as those of a cock and the bird exhibited male behaviour, attempting copulation with hens with apparent success. It was not known to crow and did not fight with cocks. It was known to be laying, but since it was suspected of laying small eggs, it was killed. When it was being prepared for the table it was noticed that the right side of the body was distinctly larger than the left and that there was a testis on the right and on the left an ovary (which really was an ovotestis) and an oviduct. The histological examination showed that it could have laid and that there were normal-looking spermatozoa and a testis of normal appearance. Every bone on the right side was larger than the corresponding bone on the left, these latter being 66 per cent. as heavy and 85 per cent. as long.

These cases can best be interpreted as gynandromorphs in a group in which the male is monogametic (XX) and in which the control of the processes of sexual differentiation is not absolutely centred in the gonads. In a male zygote one X-chromosome was eliminated at the first cleavage division of the fertilised egg, or according to the grade of femaleness in the characterisation, later at some division of a cell in those areas which are to be involved in the sexual characterisation. Evidence in support of the hypothesis that aberration in the distribution of the sex-chromosomes can occur in the somatic cells is to be obtained from a study of the abnormalities in the distribution of sex-linked plumage colour characters. Serebrovsky (1925) explains the occurrence of black feathers in a barred silver plumage by postulating that in the cells in which these feathers have their origin, the sex-chromosome, in which the sex-linked factors for barring and for silver are resident, is lost during its transference from mother to daughter cells. The question of the presence of ovotestes in these cases will be discussed later.

The frequency with which gynandromorphism is recognised in different groups varies with the interest taken by the collector or experimentalist in these groups and with the degree of sex-dimorphism that is exhibited. In many forms male and female are very similar in their external characters and in these gynandromorphism will not be detected. Moreover in many of the forms in which an occasional instance of gynandromorphism is reported (*e.g.* Arthropoda, Crustacea, Mollusca, Echinoderma) not enough is known of the mechanism which in these cases determines sex and models sexual differentiation to allow anything more than a very speculative interpretation being made.

D. BALANCED INTERSEXUALITY.

In Drosophila melanogaster.

The 2X : 3A (-IV) *male and the* 2X : 3A *female intersexes.* In an experiment made by Bridges (1921) to determine the locus of a new second chromosome recessive mutant "brown" by means of a back-cross to plexus and speck, one culture produced 96 females, 9 males, and about 80 individuals which were abnormal in their sexual characterisation. These intersexes were large-bodied, coarse-bristled flies with large roughish eyes and scolloped wing margins. The intersexes showed a bimodal variation, one group more closely approaching the female in characterisation, the other the male. All were impotent. In the female type the genitalia were predominantly or completely female, the abdomen was very much as that of the normal female; spermathecae were present. The gonads were typically rudimentary ovaries; in many cases the ovarian tissue had a bud of testicular tissue, in others one gonad was a rudimentary testis, the other a rudimentary ovary. Sexcombs were usually present. In the male type, the characterisation of the abdomen and genitalia was predominantly male. Sexcombs were always present. The gonads were typically rudimentary testes, though in a few cases the testes were well developed and contained bundles of sperm.

Moreover, instead of getting but two classes of offspring among the progeny of the back-cross as would be expected, three classes appeared, plexus speck, plexus brown, and brown speck. This unexpected result can only be explained on the understanding that

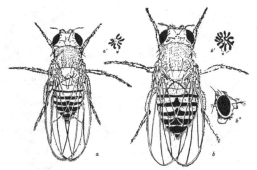

Fig. 15. (a) normal ♀ (2X : 2A); (b) triploid ♀ (3X : 3A) drawn to scale.
(*After* Morgan, Bridges and Sturtevant.)

the mothers had three instead of two II-chromosomes, one carrying plexus brown, one plexus speck, and one brown speck. From their father each had received a II-chromosome carrying plexus, brown, and speck, and from the mothers they had received one of the three kinds of II-chromosome mentioned. Each had two genes for each of the recessive characters, plexus, brown, and speck, and one for each of the alternative dominants.

Further tests showed that the mothers were also triploid in respect of chromosomes I and III, whilst chromosome IV was present either in the duplex or triplex state. It was noticed that these $3N$ (where $N =$ the haploid chromosome number) mothers could be distinguished from the ordinary normal female by their large size, coarse bristles, and rough eyes. That the intersexes were themselves triploid and not diploid with respect to chromosome II was shown by the fact that the classes, with respect to plexus, brown, and speck, among them were strikingly different from those in the sexually normal males and females. The intersexes presented three classes also but these were plexus, speck, and brown respectively.

It was readily possible to put this explanation of triploidy to the test of cytological examination and it was found that all the intersexes carried two X-chromosomes, and three of each of chromosomes II and III. Certain of them carried a Y-chromosome and some of them had three IV's, others two. It was thought at first that the triplo-IV's were the male type, the diplo-IV's the female type of

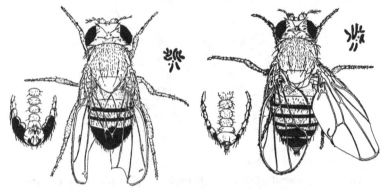

Fig. 16. Male type intersex   Fig. 17. Female type intersex
(2X : 3A-IV).   (2X : 3A).
(*Both after* Morgan, Bridges and Sturtevant.)

intersex, but further investigation has shown that this is not so; the triplo-IV's are the female intersexes, the diplo-IV's the male. It is to be noted that the infecund female type of intersex differs from her very fecund mother in that she is 2X : 3A whereas the mother is 3X : 3A, and that in these intersexes the addition of a fourth chromosome makes all the difference between a male type and a female type.

For the production of these intersexes polyploidy must have occurred in earlier generations. Bridges has shown that in certain cases ordinary diploid females possess ovaries in which there are areas the component cells of which are much larger than the normal and that in these the chromosomes are tetraploid ($4N$). Evidently in some oogonial cell there had been chromosome division of the nucleus so that all mature eggs would be diploid. The same process could occur in spermatogenesis. Should such a diploid gamete fuse with a normal haploid gamete from the mother parent it would give rise to a $3N$ zygote, so

$2X : 2A$ egg   × $1X : 1A$ sperm = $3X : 3A$ zygote = $3N$ female.
$2X : 2A$ egg   × Y : $1A$ sperm = $2X : 3A$ zygote = intersex.
XY : $2A$ sperm × $1X : 1A$ egg   = $2X : 3A$ zygote = intersex.

The following represents the series of sex types in *Drosophila melanogaster*. If the efficiency of the female determining genes (X-borne) is represented as 100 and that of the male determining gene-complex in the autosome as 80, a series of sex-indices can be made (Bridges, 1925).

Table I.

Chromosome relation	Sex types	Numerical Ratio X(100) : A(80)		Sex Index	Interval °/₀	X = -6 A = +2
$3X : 2A$	Super-female (triplo-X)	1·5 :	1	1·88	50	– 14
$4X : 4A$	$4N$ female ⎫	1 :	1	1·25	—	– 20
$3X : 3A$	$3N$ female ⎪ (normal since	1 :	1	1·25	—	– 12
$2X : 2A$	$2N$ female ⎬ ratio X:A is	1 :	1	1·25	—	– 8
$1X : 1A$*	$1N$ female ⎭ same in all)	1 :	1	1·25	50	– 4
$2X : 3A$	Intersex (female type)	1 :	1·5	0·83	—	·· 6
$2X : 3A$ (·IV)	Intersex (male type)	1 :	1·5	0·83	33	– 6
$1X : 2A$	Male (normal)	1 :	2	0·63	50	– 2
$1X : 3A$	Super-male (triplo-A)	1 :	3	0·42	—	0

* This type has only recently been described (Bridges, 1925).

A review of these facts shows that sex-determination is not the function of the sex-chromosomes alone but indicates that the initial sexuality of the zygote is determined by the interaction of the genes resident in the chromosomes, sex-chromosomes and autosomes alike. The addition of more autosomal chromatin (= autosomal-borne genes) to the usual female $2X : 2A$ balance so disturbs this that the relationship is now that which can be expressed as $2X : 3A$ and results in the establishment of a physiological state in which male type characters develop, *i.e.* the addition transforms a female type of metabolism into one which approaches that characteristic of the male. The addition of more X-chromatin (= X-borne genes) to the usual male 1X : 2A balance so disturbs this that the relationship is now that which can be expressed as 2X : 2A, one that is typical of the female.

There can be no doubt that it is far simpler for purposes of discussion to assume that in the X-chromosomes are genes that are "female-determining" and that on the autosomes are genes that are "male-determining" and that the sexuality of the zygote is determined by the balance between these. For example, if F is taken to represent a set of such female-determining genes on an X-chromosome; if M indicates a set of the male-determining genes in one set (haploid) of the autosomes, and if in their relationship 1F is greater than 1M but less than 2M, then in those classes in which the male is digametic

$$(FX)(FX)MM \text{ is a female because } 2F > 2M$$
$$(FX)YMM \quad \text{ is a male} \quad \text{ because } 1F < 2M$$

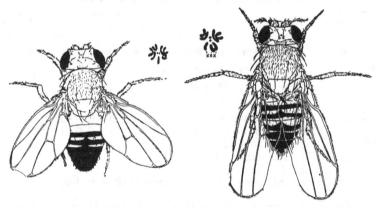

Fig. 18. A super-male (1X : 3A).   Fig. 19. A super-female (3X : 2A).
(*Both after* Morgan, Bridges and Sturtevant.)

In the classes in which the female is digametic it is necessary to place the male-determining genes on the X and the female-determining genes on the autosomes,

$$(MX)(MX)FF \text{ is a male} \quad \text{ because } 2M > 2F$$
$$(MX)YFF \quad \text{ is a female because } 1M < 2F$$

and in either case in those circumstances in which the quantitative balance between the male- and the female-determining genes is such as to lead to a situation in which M = F, intersexuality will result. This could be the situation in the ordinary hermaphrodite whilst alternating maleness and femaleness in one and the same individual might be reflections of a change-over in the relative values of F and M.

These formulae must not be taken too literally: they are but essays in convenient symbolism and it is to be understood that this is but a way of explaining an imperfectly understood subject concerning which there is a profound lack of physiological knowledge. It would appear to be an accepted fact that in *Drosophila* the effective factor in the establishment of maleness, femaleness, and intersexuality is the numerical ratio of X-chromosomes to autosomes. Underlying the above explanation is the suggestion that actual differences, to be estimated by differences in the chemical nature of the products of their functioning, undoubtedly must exist in different types of chromatin organisation (genes) and in different gene associations (chromosomes). The chromatin of the X-chromosomes is physiologically different from that in the autosomes but as yet this difference cannot be defined. At present all that can be stated is that the difference is such that more or less X-chromatin (and also of IV-chromatin) in association with more or less A-chromatin leads to one of three types of internal environment in the zygote and its cells, maleness, femaleness, or intersexuality.

In a yellow stock of *Drosophila melanogaster*, L. V. Morgan (1925) found an individual in which the two X-chromosomes were joined end to end, and from this a stock was raised. So long as this double X remains unbroken such a female will elaborate two sorts of eggs, one carrying the double X, and the other lacking an X. A double-X egg fertilised·by an X-bearing sperm will yield a triplo-X super-female. The same egg fertilised by a Y-bearing sperm will yield an XXY female. This double-X chromosome, however, is liable to break at any point along its length. Muller and Dippel (1926) have shown that by exposure to X-rays artificial breaks of this chromosome can be caused, and that under those circumstances the break is at, or near, the point of juncture of the two X's. Natural breakage results in a J-portion longer than the normal X and a shorter portion which, owing to the fact that it includes no points of attachment for the spindle, is lost. The XXY female will elaborate XX : Y : J ova. The J-bearing ovum fertilised by a Y-bearing sperm yields a normal male.

Muller and Dippel point out that during the course of the elaboration of these ova there were involved tissues which possessed an XY type of genotype like that which occurs in ordinary diploid male cells, and like the spermatogonia themselves, and they express surprise that cells of identical genetic constitution and in an identical environment may thus pursue entirely different modes of development. They point out that cells of male genetic constitution metamorphose into ova, not because of any special substance or influence exerted upon them derived from other cells of the female body, and not because of any egg-forming substances that might be present within their own protoplasm, derived from an earlier period in which they possessed the female gene complex, but because, they would suggest, the processes of oogenesis really—chemically—begin in much earlier germ cells than the oocytes and that after the early period in which oogenesis is initiated, each step in the process in turn produces the next reaction of the series in chain-like connection until the definitive ova are laid down without the further direct intervention of the genetic determiners that decided the nature (whether male or female) of the original reactions.

It will be argued herein that the form of the gamete is determined by the organisation of the gonad, this being determined by the initial general or local genotype, and that the genotype of the gamete has no relation to its form or function. In this connection it is of interest to note that both Ancel (1902–3) and Buresch (1912) have shown that the same primordial germ cells may in the Pulmonates become eggs or spermatozoa according to whether or not they enter into relations with the so-called nurse cells. As yet it cannot be stated whether the occasional eggs found amid testicular tissues in the Crustacea, Amphibia, and Mammalia are to be regarded as the products of formerly existing ovarian tissue or as the result of an abnormal differentiation of spermatogonia.

E. A "Gene" for Intersexuality.

In Drosophila simulans.

There is one outstanding example of disharmony in the geno-type that shows that there is indeed a sex-determining mechanism and that the elements of this are the genes in the chromosomes. Sturtevant (1920) has shown that there is a single recessive chromosome II gene which, when in the duplex state, turns females into intersexes and renders males sterile. Two hundred of these intersexes appeared in a stock and unlike the rather variable polyploid intersexes of *Drosophila melanogaster* were all of a definite and constant type. Their characterisation is detailed in the following table.

Table II.

	Males	Females	Intersexes
Sex combs on forelegs	Present	Absent	Absent
Number of dorsal abdominal tergites	Five	Seven	Seven
Ovipositor	Absent	Present	Present
Spermatheca	None	Two	Two
Penis	Present	Absent	Absent
First genital tergite	Present	Absent	Present
Anal plates	Lateral	Dorsal ventral	Lateral
Claspers	Present	Absent	Present
Tip of abdomen	Black	Banded	Black
Gonads	Testes	Ovaries	Very minute
Sex-chromosome constitution	XY	XX	XX (even those parts with male characterisation)

Matings between the stock that gave the intersexes and others that did not, gave no intersexes in F_1, but these appeared in F_2. Pair matings within the intersex-producing stock gave 510 females, 165 intersexes, and 754 males, an excess of males and a 3 : 1 ratio in respect of sexual normality and sexual abnormality. This character intersexuality is linked with the II-chromosome recessive character plum, for three F_1 females out of an intersex × plum mating gave in F_2:

Females		Intersexes		Males	
Wild-type	*: Plum*	*Wild-type*	*: Plum*	*Wild-type*	*: Plum*
198	91	87	0	293	65

There was no intersex plum class, showing that these characters are linked, that their genes are resident in one and the same chromosome.

These cases are of the utmost importance in that they show that intersexuality can be caused by a genetic and stable reorganisation of the chromatin in a single locus in a chromosome and that this in its effects is sufficient to transform the state of femaleness which would have become established in its absence into one of stable intersexuality. It is of interest to note that the duplex state of the gene is required to produce this result in the female, whereas in the male it leads only to a genetic type of infecundity. It should be understood that by a gene for intersexuality is meant that a certain organisation of the chromatin in a particular area of a particular chromosome, different from that organisation of the chromatin in this area that is associated with the development of a normal sexual characterisation, produces in its action, through its physiological contribution to the general economy of the zygote, a disturbance in the internal environment, so that this no longer is characterised by those properties which can be summed up as femaleness but becomes altered in the direction of maleness. This type of intersexuality is the result of a disharmony in the *composition* of the genotype and not, as was the case in balanced intersexuality, of a disharmony in its *distribution*.

It will be noticed that certain of the component structures of the sexual characterisation of the intersex are completely female and that others are completely male. There is no intermediate characterisation of parts. That this is so might conceivably be explained in one of two ways. The zygote is a genotypic female 2X : 2A in chromosome constitution but on two of the autosomes there is the gene for intersexuality. Every cell of the body will contain the same genotype. The internal environment of every cell is the same, yet some tissues differentiate according to the architectural plan of the male, others to that of the female. It may be that certain tissues are more susceptible to the stimulus of the modified environment. But another explanation and one that seems to be more reasonable is that which assumes that there is a time seriation in differentiation, those structures that assume the male characterisation becoming differentiated earlier or later than those that assume the female characterisation, and that the gene for intersexuality comes into action when one or other set has completed this differentiation. The intersexes would then be intersexes in time. The suggestion that all the component genes in a genotype do not come into action is not extravagant; it is more pleasing than that which demands a varying potency of genes. The conception that the genotype is to be regarded as an unbolting mechanism, the units of which come into action serially and evoke responses in tissues that exhibit a seriation in the time at which they can respond, will be discussed later.

F. Y-borne Genes in the Fish, and Crossing-over between the X and Y.

The results of countless breeding experiments involving sex-linkage all point to the conclusion that in *Drosophila* the Y-chromosome is empty or physiologically inert, that upon it either no genes are borne or they are inactive, unless it be granted, in the light of the non-disjunction experiments, that there is a gene for fertility on the Y, as an XO male is sterile while an XY is fertile. But the impossibility of testing this gene for fertility against a mutated state producing relative or absolute infertility, makes it impossible yet awhile to regard the suggestion seriously. XO males, XXY and XXYY females show no somatic differences from ordinary XY or XX individuals respectively, whereas haplo-IV, triplo-IV, and triplo-X individuals do show such differences. Neither deficiency nor excess of Y produces any upset in the balance of the genes so that the Y can in this connection be disregarded, though Stern (1926) has recently brought forward evidence which points to the conclusion that there are genes on the Y of *Drosophila*. The work of Schmidt (1920) and of Winge (1922) on the millions fish, *Lebistes reticulatus*, and of Aida (1921) on *Aplocheilus latipes*, the small ornamental fish of Japan, has also shown that there certainly are Y-borne genes in other forms. Schmidt and Winge found that the male and female of *Lebistes* differed markedly in appearance, the female being larger than the male and of a plain grey-green colour, while the male, according to the race to which it belongs, is adorned with red or yellow spots on the dorsal fin. The males of different races are easily distinguished, but the females of all races look alike. A male of any race mated with a female of the same or of any other race produces plain grey-green daughters and sons ornamented like himself. In the F_2 generation again all the males are coloured like the P_1 males; there is no segregation. An F_1 male mated to a female of any race produces sons like himself; an F_1 female mated to any male will produce sons coloured like their father. The mother in no way influences the colouring of her sons which persistently reproduce their father's colour. Certain characters in *Lebistes* are limited to the male. Such a mode of inheritance can readily be understood if the characters concerned are sex-linked and if their genes are resident upon the Y-chromosome in a case in which the male is digametic, and if the characters are epistatic to those (affecting the same parts) the genes for which are borne upon the autosomes that the zygote receives from the maternal parent.

Winge has shown that there are 46 chromosomes in both male and female and that it is not possible to identify the different pairs. The results of experimental breeding work demand for their interpretation that the Y-chromosome shall always bear genes for some colour or other and shall be transmitted only from father to son. Even if the X-chromosome also bears genes for colour, the female remains uncoloured, but in the male the colour is a compromise between the usual sex-linked and the sex-limited characterisations and the two kinds of sex-associated characters can be identified. As a result of his breeding experiments, Winge was able to distinguish the following sorts of X- and Y-chromosomes according to the different genes borne thereupon:

X_0 Which does not involve any colour pattern of the male.

X_s *Sulphureus*, bearing genes producing a sulphur-yellow colour in the dorsal fin, in the tail, and the caudal fin, and red colour in the lower edge of the caudal fin.

Y_r *Ruber*, bearing genes producing red colour in upper edge of caudal fin, a large oblong red spot below and behind the dorsal fin, and a dark side dot on the tail.

Y_1 *Iridescens*, producing a mother-of-pearl body sheen, two to three red side spots and black side dots on tail and body.

Y_m *Maculatus*, producing a large black dot on dorsal fin, a large red side spot, and a black dot at the gat.

Y_f *Ferrugineus*, producing black rust-coloured part in caudal fin, and a black side dot on the tail.

In one experiment in which an X_0X_0 female was crossed with an X_8Y_r male, 44 males with the constitution X_0Y_r were produced all exhibiting the characters of their male parent, red colour proximally in the upper edge of the caudal fin, a long red side spot placed below and behind the dorsal fin, a dark side spot on the tail near the caudal fin, and a colourless dorsal fin, characters the genes for

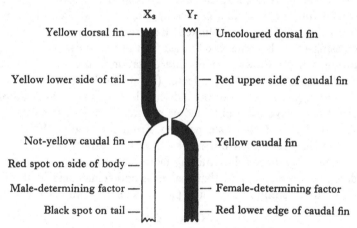

Fig. 20. Crossing-over between X and Y in *Lebistes reticulatus*.

which are resident in the Y-chromosome. But in addition there was a male which had a dorsal fin yellow in colour and yellow pigment in the lower side of the tail, characters the genes for which are resident in the X-chromosome, while it lacked completely the red and yellow colours on the caudal fin which was quite un-pigmented. This case is interpreted by Winge as an instance of crossing-over between the X_8 and Y_r chromosomes during the spermatogenesis of the X_8Y_r parent; he points out that apparently there is no more difficulty in crossing-over between the X and Y in *Lebistes* than in the case of the autosomes; all are morphologically alike.

These colour genes were not resident in the X_0 derived from the mother: all the characters based upon the Y-borne genes derived from the father were exhibited, and in addition there were characters the genes for which are borne upon the X of an X_s race. This kind of X was in association with the Y_r chromosome in the X_sY_r parent (the father). It follows, therefore, that the Y_r chromosome was not completely a Y-chromosome but was an association of part of an X_s and Y_r; such an association could be established by a crossing-over with the result that four sorts of sperm would be produced: X_s, Y_r, part X_s and part Y_r, part Y_r, and part X_s.

Winge (1923) has shown more recently that the X-chromosome may carry another gene, *elongatus* (e), responsible for an elongated caudal fin, and that this gene may cross over so as to be borne upon the Y, so that an X-linked character may, as a result of such crossing-over, become sex-limited in the transmission. The following schema shows Winge's interpretation of the case. The male used ought to have had the factor (m) in the Y-chromosome and the factor (e) in the X, but the individuals A, B, and C, for example, could only have been obtained as a result of crossing-over of the X and the Y of the male parent. From his results Winge also concluded that the chromosomes X and Y respectively contain a recessive female sex-determining factor and a dominant male sex-determining factor, and that maleness and femaleness in the fish constitute a simple Mendelian pair of allelomorphic characters.

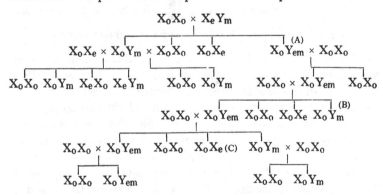

Winge termed this peculiar type of inheritance "one-sided masculine inheritance." It would seem better to reserve the term "sex-linked" for those characters the genes for which are resident in the sex-chromosomes, the X and the Y, and further to distinguish between them by using the terms X-linked to define the mode of inheritance met with in *Drosophila*, and Y-linked to define the sex-limited inheritance of characters the genes for which are resident in the Y-chromosome. Such sex-limited characters, in the absence of crossing-over, will only appear in the digametic sex.

Aida's (1921) work on *Aplocheilus* supports Winge's hypothesis and affords weighty evidence that in the fish crossing-over between X- and Y-chromosomes can and does occur. *Aplocheilus* as found in nature is brownish in colour, but four coloured varieties are known to the fancier—red, variegated red, white, and variegated white, and of these all the whites without exception are females. White is recessive to any other colour and breeds true; brown is dominant to red, giving a 3 : 1 ratio in the F_2. Brown and white gives in F_2 brown, blue (a new colour variety), red and white in the ratio 9 : 3 : 3 : 1. Blue × red give all browns. Variegated red × white give all variegated reds, and in F_2 variegated reds, variegated whites, reds, and whites in the proportions of 9 : 3 : 3 : 1. From the above it follows that in the characterisation of the skin three genes are involved.

B able to produce black pigment (melanophore) uniformly.
R   „   red (yellow) (xanthopore) uniformly.
B′   „   black pigment partially, resulting in variegated.

The genetic constitution of the different varieties will then be as follows:

Brown	BBRR	White	bbrr
Blue	BBrr	Variegated white	B′B′rr
Red	bbRR	Variegated red	B′B′RR

Aida found that when a white female was mated to a brown, red, or variegated red male (*i.e.* any male duplex for R), all the white individuals in F_2 were females (with a few exceptions). On the other hand, having bred white males and mated them to females duplex for R, *e.g.* BBRR, bbRR, B′B′RR, he found that in the F_2 all the whites, blues, and variegated whites were males (again with a few exceptions). These extraordinary results Aida interpreted as follows: (1) In this fish the male possesses the XY type of sex-organisation. (2) The gene for red is located in the chromosomes X and Y. A homozygous red male will have the constitution bb(RX)(RY), a female duplex for R will be bb(RX)(RX). In no other manner can the production of red males (red being dominant) from the cross red male × white female be explained.

<pre>
      Red ♂   White ♀
    bb(RX)(RY)   ×   bb(rX)(rX)
      /   \ |
   b(RX) b(RY)   b(rX)   Gametes

      bb(RX)(rX)   bb(rX)(RY)   ×   bb(rX)(rX)
       Red ♀♀   Red ♂♂   White ♀♀
                        / \ |
                    b(rX) b(RY)   b(rX)   Gametes

                    bb(rX)(rX)   bb(rX)(RY)
                     White ♀♀   Red ♂♂
</pre>

The Y-chromosome is restricted to the male line, so that all males in the progeny of any mating in which either a brown BB(RX)(RY) or a red bb(RX)(RY) male is used must be red, variegated red, or brown, whereas all individuals lacking this Y-borne dominant gene R—whites, variegated whites, and blues— must be females.

Aida found further that evidence of crossing-over between the X- and Y-chromosomes was not rare in *Aplocheilus*: for example, in the heterozygous red male × white female experiment in which the females produced were white and the males red, exceptional red females appeared having the constitution bb(RX)(rX), and exceptional white males, bb(rX)(rY), were also occasionally encountered. Aida satisfied himself by further breeding that these were not instances of the results of non-disjunction in which case the exceptional red females would have the constitution bb(rX)(rX)(RY) and the exceptional white males, bb(rX)O. To test this question as to which of the two respective constitutions the exceptional male individuals had, the following experiment was carried out:

<pre>
   bb(RX)(rX)   ×   bb(RX)(rY)   or   bb(RX)O
  Heterozygous   Heterozygous   Non-disjunctional
                    exceptional   Red ♂
                     Red ♂
</pre>

The offspring of the mating were as follows:

<pre>
Red : White = 3 : 1   Reds   146 ♀♀ : 57 ♂♂
                      Whites   2 ♀♀ : 80 ♂♂
</pre>

The fact that two white females were produced shows conclusively that the gametic constitution of the red male could not be bb(RX)O, since the white female must be bb(rX)(rX) and in the mating bb(RX)(rX)×bb(RX)O there is but one single gene r, thus:

$$
\begin{array}{ccccc}
\text{bb(RX)(rX)} & & \times & & \text{bb(RX)O} \\
\text{b(RX)} & \text{b(rX)} & & \text{b(RX)} & \text{bO} \\
\text{bb(RX)(RX)} & \text{bb(RX)O} & & \text{bb(RX)(rX)} & \text{bb(rX)O}
\end{array}
$$

and no white females.

Whereas, if the male had the other constitution bb(RX)(rY) and if a crossing-over occurred between the chromosomes RX and rY of the male, so that (rX) and (RY) resulted, the production of bb(rX)(rX) females may be explained clearly:

$$
\begin{array}{ccccc}
\text{bb(RX)(rX)} & \times & & \text{bb(RX)(rY)} & \\
\text{b(RX)} \quad \text{b(rX)} & & \text{b(RX)} & \text{b(rY)} \quad \text{b(rX)} & \text{b(RY)} \\
\end{array}
$$

(Cross-over gametes)

bb(RX)(RX)	bb(RX)(rY)	bb(RX)(rX)	bb(RX)(RY)
Red ♀	Red ♂	Red ♀	Red ♂
bb(RX)(rX)	bb(rX)(rY)	bb(rX)(rX)	bb(rX)(RY)
Red ♀	White ♂	White ♀	Red ♂
		Exceptional	

In *Drosophila*, however, there are sufficient reasons for holding that the Y-chromosome contains no genes that affect sex-determination. It follows then, that in the case of the female tissues there are two X-chromosomes in association with two of each of the autosomes, so that if A represents one complete set of autosomes as present in the ripe egg, the quantitative relation, in the body cells of the female, in the gametes, and in the XX type of zygote, between the chromatin in the sex-chromosomes and that in the autosomes can be expressed by the formula 1X : 1A, whereas in the male somatic cells and immature gametes, and in the XY type of zygote it is expressed by the formula 1X : 2A. In the X-chromosome-bearing sperm the relation is 1X : 1A, in the Y-bearing oX : 1A. The fact that the egg and one kind of sperm are both, in this respect, 1X : 1A, and that the other kind of sperm is oX : 1A is of interest in that it shows quite conclusively that gamete and zygote are entirely dissimilar in spite of the fact that each may exhibit a similar X : A ratio. In non-disjunction it is common to get an XY-bearing egg which is, in every way, an egg, and in no way an XY individual, a male. That part of the geno-type which is concerned in the morphological aspects of sexuality in the zygote is not liberated until egg has been activated by sperm. As will be seen, the genotype of the gamete has no constant relation to the structure of the gamete, for the latter is determined by the kind of organisation of the gonad that manufactured it: an ovary elaborates eggs without reference to their genotype and a testis elaborates sperm without reference to their genotype. A gamete 1X : 1A in genotype will be an egg if it is the elaborated product of an ovary; a sperm if it is elaborated by a testis.

Since it is established that the genes on the X-chromosome are different, as judged by the end-results of their action, from those on the autosomes, and that the chromatin organisation of any chromosome can exhibit change, it follows that chromatin of the X and of the autosomes is different in its organisation and there-fore in its physiological activity. Thus the quantitative balance between the X-borne and the autosomal-borne genes differs in male and female tissues and the physiological states established by the interaction of genes in male and female zygotes respectively will likewise be dissimilar. The conception of contrasted sex-determining factors and substances was suggested by Morgan (1911, 1913), and it was Woltereck (1911) who suggested that these latter might be of the nature of zymogens. As will be seen later, this idea has been greatly elaborated by Goldschmidt.

It is not necessary—it is probably entirely incorrect, though it is undoubtedly convenient—to postulate male-determining and female-determining genes especially concerned in determining sexuality and elaborating "andrase" and "gynase," the male-differentiating and female-differentiating substances respectively. Maleness in Drosophila is that physiological state or quality established in a zygote as the result of the action and interaction of all the genes in a genotype that may be symbolised as 1X : 2A. In the beginning maleness in the zygote is a certain kind of internal environment, a certain metabolic level; femaleness is another, and in one or the other organogeny and ontogeny proceed. In the XX type of individual an internal environment of femaleness becomes expressed and the structures that pertain to the sexual organisation will pursue their differentiation under the direction of their own genotype, in the presence of the physiological stimuli exerted by other differentiating and differentiated tissues, and in an external environment which may condition this differentiation. On the other hand, in the XY zygote an initial internal environment of maleness becomes established and the differentiation of the sexual organisation during ontogeny will be such as will yield a typical phenotypic male.

An individual in the case of *Drosophila* is a phenotypic male (1) because in the beginning it was a genotypic male (XY); (2) because in this XY zygote the relation between the X-borne and the autosomal-borne genes (the ratio between their number or their algebraic sum, as suggested by Schrader and Sturtevant (1923) as an explanation of the facts of sexuality in those cases in which one sex is haploid, the other diploid), was such that an internal environment of maleness became established; (3) because in this internal environment the structures concerned in the sexual organisation, being XY in chromosome constitution, developed, and in developing flourished; and (4) because the impress of the external environment did not or could not affect the developing zygote in such a way as to override the effects of the gene mechanism and to modify the internal environment based thereupon so that it would become physiologically equivalent to the type of internal environment normally associated with the genotype symbolised as XX. For similar reasons the genotypic female becomes a phenotypic female. *Propter secretiones internas totas mulier est quod est*, a female is a female by the totality of her internal secretions, remains true only if it be granted that the phrase internal secretions shall include the metabolic products of her initial genotype.

Sex, physiologically, is an equitable division of labour between two kinds of individuals within a (bisexual) species, one of these being anabolic, the other catabolic, and this difference in the rate and degree of the processes of metabolism is exhibited in every activity of the individual, by the individual cell as well as by the body as a whole. The sex dimorphic characters are the end results of the differential development of a common series of anlagen, for sexuality implies the conception of reciprocal differences, maleness and femaleness being but two dissimilar manifestations of a common series of structures and functions.

56   MECHANISM OF SEX-DETERMINATION

BIBLIOGRAPHY (A)

AGAR, W. E. (1924). Experiments with Certain Plumage Colour and Pattern Factors. *Jour. Genet.* 14, pp. 265–272.

AIDA, T. (1921). On the Inheritance of Colour in a Fresh-water Fish, *Aplocheilus latipes*, with special reference to Sex-linked Inheritance. *Genet.* 6, pp. 554–573.

ALLEN, EZRA (1918). Studies on Cell Division in the Albino Rat. III. Spermatogenesis. *Jour. Morph.* 31, pp. 133–186.

ANCEL, P. (1902–3). Histogénèse et structure de la glande hermaphrodite d'*Helix pomatia*. *Arch. d. biol.* 19, pp. 389–652.

BACHHUBER, L. J. (1916). The Behaviour of the Accessory Chromosomes and of the Chromatid Body in the Spermatogenesis of the Rabbit. *Biol. Bull.* 30, pp. 294–310.

BAEHER, V. W. B. (1908). Ueber die Bildung von Sexualzellen bei Aphididae. *Zool. Anz.* 33, pp. 507–517.

—— (1909). Die Oogenese bei einigen viviparen Aphiden und die Spermatogenese von *Aphis saliceti*. *Arch. f. Zellf.* 3, pp. 269–334.

—— (1911). Contribution à l'étude de la cariocinèse somatique, de la pseudo-réduction et de la réduction. *La Cellule*, 27, pp. 385–450.

BAKER, J. R. (1926). Sex in Animals and Man. London. 175 pp.

BALTZER, F. (1913). Ueber die Herkunft der Idiochromosomen bei Seeigeln. *Sitzb. phys.-med. Ges. Würz.* 6, p. 90.

BANTA, A. M. AND BROWN, L. A. (1924). Rate of Metabolism and Sex Determination in Cladocera. *Proc. Soc. Exp. Biol. Med.* 22, pp. 77–79.

BAUMGARTNER, W. J. (1902). Spermatid Transformation in *Gryllus*. *Kans. Univ. Sci. Bull.* No. 1, pp. 47–63.

—— (1904). Some New Evidence for the Individuality of the Chromosomes. *Biol. Bull.* 8, pp. 1–28.

BERRY, H. E. (1906). The "Accessory Chromosome" in Epeira. *Biol. Bull.* 11, pp. 193–201.

BLACHER, L. J. (1926). The Influence of Sexual Hormones on the Number of Red Corpuscles and Haemoglobin Percentage in the Blood of the Fowl. *Trans. Lab. Exp. Biol., Moscow*, 5, pp. 9–17.

—— (1926). The Effect of Testes on Phosphorus Metabolism. *Ibid.*

—— (1926). The Influence of the Gonad upon Male Sexual Characters in *Lebistes reticulatus*. II. A Case of Hermaphroditism. *Ibid.* 1, pp. 90–95.

BLACKMAN, M. W. (1903). On the Chromatin in the Spermatocytes of *Scolopendra heros*. *Biol. Bull.* 5, pp. 187–217.

—— (1905). The Spermatogenesis of *Scolopendra heros*. *Bull. Mus. Comp. Zool. Harv.* 48, pp. 1–141.

—— (1910). Spermatogenesis of the Myriapods. VI. *Biol. Bull.* 19, pp. 138–159.

BOND, C. J. (1913). On a Case of Unilateral Development of Secondary Male Characters in a Pheasant, with Remarks on the Influence of Hormones in the Production of Secondary Sex Characters. *Jour. Genet.* 3, pp. 205–217

BORING, A. M. (1907). A Study of the Spermatogenesis of Twenty-two Species of the Membracidae, Jassidae, Cercopidae and Fulgoridae. *Jour. Exp. Zool.* 4, pp. 469–513.

—— (1913). The Odd Chromosome in *Cerastipsocus venosus*. *Biol. Bull.* 24, pp. 125–128.

BOVERI, T. (1911). Ueber das Verhalten der Geschlechtschromosomen bei Hermaphroditismus. Beobachtungen an *Rhabditis nigrovenosa*. *Verh. phys.-med. Ges. Würzb.* 41, pp. 83–97.

—— (1915). Ueber die Entstehung der Eugsterschen Zwitterbienen. *Arch. f. Entw.* 41, pp. 264–311.

BRIDGES, C. B. (1916). Non-disjunction as a Proof of the Chromosome Theory of Heredity. *Genet.* 1, pp. 1–62, 107–163.

—— (1919). Vermilion-deficiency. *Jour. gen. Physiol.* 1, pp. 645–656.

—— (1921). Triploid Intersexes in *Drosophila melanogaster*. *Sci.* 54, pp. 252–254.

—— (1925). Sex in Relation to Chromosomes and Genes. *Amer. Nat.* 59, pp. 127–137.

—— (1925). Haploidy in *Drosophila melanogaster*. *Proc. Nat. Acad. Sci.* 11, pp. 706–710.

BROWNE, E. (1910). The Relation between Chromosome Number and Species in *Notonecta*. *Biol. Bull.* 20, pp. 19–34.

BURESCH, F. (1912). Untersuchungen über die Zwitterdrüse der Pulmonaten. *Arch. f. Zellf.* 7, pp. 314–343.

BÜTSCHLI, O. (1876). Studien über die ersten Entwicklungsvorgänge der Eizelle, die Zellteilung und die Konjugation bei Infusorien. *Abh. Senkenb. naturf. Ges.* 10, pp. 213–464.

CALKINS, G. N. (1904). Studies on the Life History of Protozoa. *Jour. Exp. Zool.* 1, pp. 423–462.

CHARLTON, H. H. (1921). The Spermatogenesis of *Lepisma domestica*. *Jour. Morph.* 35, pp. 381–423.

CHICKERING, A. M. (1918). Chromosomes of *Ranatra* sp.? *Trans. Amer. Micrs. Soc.* 37, pp. 132–133.

CORRENS, C. AND GOLDSCHMIDT, R. (1913). Die Vererbung und Bestimmung des Geschlechts. Berlin.

CUNNINGHAM, J. T. (1900). Sexual Dimorphism in the Animal Kingdom. London.

DAVIES, H. S. (1908). Spermatogenesis in Acididae and Locustidae. *Bull. Mus. Comp. Zool. Harv.* 53, pp. 57–158.

DONCASTER, L. (1908). Sex-Inheritance in the Moth *Abraxas grossulariata* and its Variation *Lacticolor*. *4th Rept. Evol. Com.* pp. 53–57.

—— (1914). On the Relation between Chromosomes, Sex-limited Transmission and Sex Determination in *Abraxas grossulariata*. *Jour. Genet.* 4, pp. 1–22.

—— (1914). *The Determination of Sex*. Cambridge.

DZIERZON, J. (1845). *Eichstädter Bienenzeit.* p. 252.

EDLBACHER, S. AND RÖTHLER, H. (1925). Beiträge zur Kenntniss der Arginase. III. Argininumsatz und Sexualität. *Zeit. physiol. Chemie,* 148, pp. 273–282.

58   MECHANISM OF SEX-DETERMINATION

EDWARDS, C. L. (1910). The Idiochromosomes of *Ascaris megalocephala* and *A. lumbricoides*. *Arch. f. Zellf.* 5, pp. 422–429.
—— (1912). The Sex Chromosomes of *A. felis*. *Ibid.* 7, pp. 309–313.
ELLIS, HAVELOCK (1914). *Man and Woman.* Walter Scott.
ENGELHARDT (1914). Ueber den Bau der gynandromorphen Bienen. *Zeit. wiss. Insektenbiol.*
FOOT, C. AND STROBELL, E. C. (1912). A Study of Chromosomes and Chromatin Nucleoli in *Euschistus*. *Arch. f. Zellf.* 9, pp. 47–62.
GEDDES, P. AND THOMSON, J. A. (1899). *The Evolution of Sex.* London.
GEINITZ, B. (1915). Ueber Abweichungen bei der Eireifung von *Ascaris*. *Arch. f. Zellf.* 13, pp. 588–630.
GOLDSCHMIDT, R. (1923). *The Mechanism and Physiology of Sex Determination.* London.
GOODRICH, E. S. (1916). The Germ Cells of *Ascaris incurva*. *Jour. Exp. Zool.* 21, pp. 61–100.
GOSIO, B. (1905). Zur Methodik der Pestvaccin-Bereitung. *Zeit. f. Hyg.* 50, pp. 519–528.
GRÄFENBERG, E. (1922). Die Geschlechtsspezifizität des weiblichen Blutes. *Arch. f. Gynäk.* 117, pp. 52–55.
GRANATA, L. (1910). Le cinesi spermatogenetiche di *Pamphagus marmoratus*. *Arch. f. Zellf.* 5, pp. 183–214.
GROSS, J. (1904). Die Spermatogenese von Syromastes. *Zool. Jahrb.* 20, pp. 439–498.
GULICK, A. (1911). Ueber die Geschlechtschromosomen bei einigen Nematoden. *Arch. f. Zellf.* 6, pp. 339–382.
GUTHERZ, S. (1906–7). Zur Kenntniss der Heterochromosomen. *Arch. mikr. Anat.* 69, pp. 491–514.
HALDANE, J. B. S. (1921). Sex-linked Inheritance in Poultry. *Sci.* 54, p. 663.
HALDANE, J. B. S. AND CREW, F. A. E. (1925). Change of Linkage in Poultry with Age. *Nat.* 115, p. 641.
HARVEY, E. B. (1916). A Review of the Chromosome Number in the Metazoa. *Jour. Morph.* 28, pp. 1–64.
—— (1920). II. *Ibid.* 34, pp. 1–68.
HENKING, H. (1890). Reduktionstheilung der Chromosomen in den Samenzellen von Insekten. *Intern. Monatsh. f. Anat. Physiol.* 7, p. 243.
HOGBEN, L. T. (1920). Studies on Synapsis. II. Parallel Conjugation and the Prophase Complex in Periplaneta, with Special Reference to the Premiotic Telophase. *Proc. Roy. Soc. B*, 91, pp. 305–329.
HUETTNER, A. (1922). The Origin of the Germ-Cells in *Drosophila melanogaster*. *Jour. Morph.* 37, pp. 385–424.
HUXLEY, J. S. (1923). Courtship Activities of the Red-throated Diver (*Colymbus stellatus* Pontopp.) together with a Discussion of the Evolution of Courtship in Birds. *Linn. Soc. Jour.* 35, pp. 253–292.
JENNINGS, H. S. (1913). The Effect of Conjugation in Paramecium. *Jour. Exp. Zool.* 14, pp. 279–391.

JORDAN, H. E. (1908). The Spermatogenesis of *Aplopus mayeri*. *Carn. Inst. Publ.* No. 102, pp. 15–38.
—— (1913). A Comparative Study of Mammalian Spermatogenesis with Special Reference to Heterochromosomes. *Sci.* 37, pp. 270–271.
KAUTSCH, G. (1913). Studien über Entwicklungsanomalien bei Ascaris. II. *Arch. f. Entw.* 35, pp. 642–691.
KORNHAUSER, S. I. (1914). A Comparative Study of the Chromosomes in the Spermatogenesis of *Euchenopa binotata* and *E. curvata*. *Arch. f. Zellf.* 12, pp. 241–298.
KYBER, J. F. (1815). Einige Erfahrungen und Bemerkungen über Blattläuse. *Germars Mag. Entom.*
LAWRENCE, J. V. AND RIDDLE, O. (1916). Studies in the Physiology of Reproduction in Birds. VI. Sexual Differences in the Fat and Phosphorus in the Blood of Fowls. *Amer. Jour. Physiol.* 41, pp. 430–437.
LEFÈVRE, G. AND McGILL, C. (1908). The Chromosomes of *Anasa tristis* and *Anax junius*. *Amer. Jour. Anat.* 7, pp. 469–487.
LIPSCHÜTZ, A. (1924). *The Internal Secretion of the Sex Glands*. Cambridge.
LITTLE, C. C. (1920). Is the Foetal Tortoiseshell Tom-cat a Modified Female? *Jour. Genet.* 10, pp. 301–302.
LOEB, J. (1918). The Sex of Parthenogenetic Frogs. *Proc. Nat. Acad. Sci.* 3, pp. 60–62.
McCLUNG, E. C. (1901). Notes on the Accessory Chromosome. *Anat. Anz.* 20, pp. 220–226.
—— (1902). The Accessory Chromosome—Sex Determinant? *Biol. Bull.* 3, pp. 43–94.
MACKLIN, M. (1923). Description of Material from a Gynandromorph Fowl. *Jour. Exp. Zool.* 38, pp. 355–375.
MALONE, J. Y. (1918). Spermatogenesis in the Dog. *Trans. Amer. Micr. Soc.* 37, pp. 97–110.
MANOILOV, E. O. (1922–3). Identification of Sex in Plants by Chemical Reaction. *Bull. App. Bot.* (Russian), 13, pp. 503–505.
MARCHAL, E. AND E. (1906). Recherches expérimentales sur la sexualité des spores chez les mousses dioïques. *Mém. Acad. Roy. Belg.* N.S., 1.
—— —— (1907). Aposporie et sexualité chez les mousses. *Bull. Acad. Roy. Belg., Classe d. sci.* pp. 765–789.
MAUPAS, E. (1888). Recherches expérimentales sur la multiplication des infusoires ciliés. *Arch. zool. exp. gén.* 6, pp. 165–277.
—— (1889). Le rajeunissement karyogamique chez les ciliés. *Ibid.* 7, pp. 149–217.
—— (1890). Sur la multiplication et la fécondation d'*Hydatina senta* Ehr. *C. R. Acad. Sci.* 111, pp. 310–312.
—— (1891). Sur la détermination de la sexualité chez l'*Hydatina senta*. *Ibid.* 113, pp. 388–390.
MEDES, G. (1905). The Spermatogenesis of *Scutigera forceps*. *Biol. Bull.* 9, pp. 156–177.

60 MECHANISM OF SEX-DETERMINATION

MEEK, C. F. U. (1913). The Problem of Mitosis. *Q.J.M.S.* 58, pp. 567–592.

MEHLING, E. (1915). Ueber die gynandromorphen Bienen des Eugsterschen Stockes. *Verh. phys.-med. Ges. Würzb.*

MEISENHEIMER, J. (1922). *Geschlecht und Geschlechter im Tierreich.* Jena.

MITCHELL, C. W. (1913). Experimentally Induced Transition in the Morphological Characters of *Asplanchna amphora* Hudson, together with Remarks on Sexual Reproduction. *Jour. Exp. Zool.* 15, pp. 91–130.

—— (1913). Sex-determination in *Asplanchna amphora. Ibid.* pp. 225–254.

MOHR, O. L. (1915). Sind die Heterochromosomen wahre Chromosomen? *Arch. f. Zellf.* 14, pp. 151–176.

MONTGOMERY, T. H. (1901). A Study of Germ-Cells of Metazoa. *Trans. Amer. Phil. Soc.* 20, pp. 154–236.

—— (1906). Chromosomes in the Spermatogenesis of Hemiptera Heteroptera. *Ibid.* 21, pp. 97–174.

MORGAN, L. V. (1925). Polyploidy in *Drosophila melanogaster* with two Attached X-chromosomes. *Genet.* 10, pp. 148–178.

MORGAN, T. H. (1908). The Production of Two Kinds of Spermatozoa in Phylloxerans. *Proc. Soc. Exp. Biol. Med.* 5, pp. 56–57.

—— (1909). Sex Determination and Parthenogenesis in Phylloxerans and Aphids. *Sci.* 29, pp. 234–237.

—— (1909). A Biological and Cytological Study of Sex Determination in Phylloxerans and Aphids. *Jour. Exp. Zool.* 7, pp. 239–352.

—— (1911). An Attempt to Analyse the Constitution of the Chromosomes on the Basis of Sex-limited Inheritance in *Drosophila. Ibid.* 11, pp. 365–411.

—— (1912). The Elimination of Sex Chromosomes from the Male Producing Eggs of Phylloxerans. *Ibid.* 12, pp. 479–498.

—— (1913). Factors and Unit Characters in Mendelian Heredity. *Amer. Nat.* 47, pp. 5–16.

—— (1914). *Heredity and Sex.* Columbia University Press.

—— (1915). The Predetermination of Sex in Phylloxerans and Aphids. *Jour. Exp. Zool.* 19, pp. 285–315.

—— AND BRIDGES, C. B. (1919). The Origin of Gynandromorphs. *Carn. Inst. Publ.* No. 278, pp. 1–262.

MORGAN, T. H., BRIDGES, C. B. AND STURTEVANT, A. H. (1925). The Genetics of *Drosophila. Bibliog. Genet.* 2, pp. 1–262.

MORSE, M. (1909). The Nuclear Components of the Sex Cells of Four Species of Cockroaches. *Arch. f. Zellf.* 3, pp. 485–521.

MULLER, H. J. AND DIPPEL, A. L. (1926). Chromosome Breakage by X-rays and the Production of Eggs from Genetically Male Tissue in *Drosophila. Brit. Jour. Exp. Biol.* 3, pp. 85–122.

MÜLSOW, W. (1912). Der Chromosomenzyklus bei *Ancyracanthys cystidicola. Arch. f. Zellf.* 9, pp. 63–72.

NACHTSHEIM, H. (1913). Cytologische Studien über die Geschlechtsbestimmung bei der Honigbiene. *Arch. f. Zellf.* 11, pp. 169–241.

NACHTSHEIM, H. (1923). Parthenogenese, Gynandromorphismus und Geschlechtsbestimmung bei Phasmiden. *Zeit. indukt. Abst.* **30**, pp. 287–289.

NEWELL, W. (1914). Inheritance in the Honey Bee. *Sci.* **41**, pp. 218–219.

NEWMAN, H. H. AND PATTERSON, J. T. (1909). A Case of Normal Identical Quadruplets in the Nine-banded Armadillo. *Biol. Bull.* **17**, pp. 181–188.

NUSSBAUM, M. (1897). Die Entstehung des Geschlechts bei *Hydatina senta*. *Arch. mikr. Anat.* **49**, pp. 227–308.

OETTINGER, R. (1909). Samenreifung und Samenbildung bei *Pachyiulus*. *Arch. f. Zellf.* **3**, pp. 563–626.

OGUMA, K. (1921). The Idiochromosomes of the Mantis. *Jour. Coll. Agric. Hokk. Imp. Univ.* **10**, pp. 1–27.

PAINTER, T. S. (1921). Studies in Reptilian Spermatogenesis. I. On the Spermatogenesis of Lizards. *Jour. Exp. Zool.* **34**, pp. 281–328.

—— (1922). Studies in Mammalian Spermatogenesis. I. The Spermatogenesis of the Opossum. *Ibid.* **35**, pp. 13–45.

—— (1923). II. The Spermatogenesis of Man. *Ibid.* **37**, pp. 291–321.

—— (1924). III. The Fate of the Chromatin Nucleolus in the Opossum. *Ibid.* **39**, pp. 197–248.

—— (1924). IV. The Sex Chromosomes of Monkeys. *Ibid.* pp. 433–451.

—— (1924). V. The Chromosomes of the Horse. *Ibid.* pp. 229–248.

—— (1925). Chromosome Numbers in Mammals. *Sci.* **61**, pp. 423–424.

PARKES, A. S. (1923). Head-length Dimorphism of Mammalian Spermatozoa. *Q.J.M.S.* **67**, pp. 617–625.

PARMENTER, C. L. (1919). Chromosome Number and Pairs in the Somatic Mitoses of *Amblystoma tigrinum*. *Jour. Morph.* **33**, pp. 169–249.

PAYNE, F. (1908). On the Sexual Differences of the Chromosome Groups in *Galgulus oculatus*. *Biol. Bull.* **14**, pp. 297–303.

—— (1909). Some New Types of Chromosome Distribution and their Relation to Sex. *Ibid.* **16**, pp. 119–168.

—— (1910). The Spermatogenesis of *Acholla multispinosa*. *Ibid.* **18**, pp. 174–179.

—— (1912). A Further Study of the Chromosomes of the Reduviidae. II. *Jour. Morph.* **23**, pp. 331–348.

POLL, H. (1909). Zur Lehre von sekundären Sexualcharakteren. *Sitzb. Ges. naturf. Freun. Berlin*, **6**, pp. 331–358.

PROWAZEK, S. (1900–2). Spermatologische Studien. II. Spermatogenese des Nashornkäfers (Oryctes). *Arb. zool. Inst. Wien*, **13**, pp. 223–236.

PUNNETT, R. C. (1903). On Nutrition and Sex-Determination in Man. *Camb. Phil. Soc. Trans.* **12**, pp. 262–267.

ROBERTSON, W. R. B. (1908). The Chromosomes Complex of *Syrbula admirabilis*. *Kans. Univ. Sci. Bull.* **4**, pp. 275–305.

—— (1915). Chromosome Studies. III. Inequalities and Deficiencies in Homologous Chromosomes. *Jour. Morph.* **26**, pp. 109–141.

SATINA, S. AND BLAKESLEE, A. F. (1925). Studies on Biochemical Differences between (+) and (−) Sexes in Mucors. *Proc. Nat. Acad. Sci.* **11**, pp. 528–534.

62   MECHANISM OF SEX-DETERMINATION

SATINA, S. AND DEMEREC, M. (1925). Manoilov's Reaction for Identification of the Sexes. *Sci.* **62**, pp. 225–226.

SCHELLENBERG, A. (1913). Das accessorische Chromosom in den Samenzellen der Locustidae *Diestrammena marmorata* de Hahn. *Arch. f. Zellf.* **11**, pp. 399–514.

SCHLEIP, W. (1911). Das Verhalten des Chromatin bei *Angiostomum (Rhabditis) nigrovenosum. Arch. f. Zellf.* **7**, pp. 87–137.

SCHMIDT, J. (1920). The Genetic Behaviour of a Secondary Sexual Character. *C. R. d. trav. labor. Carlsberg*, **14**, 12 pp.

SCHRADER, F. (1921). The Chromosomes of *Pseudococcus nipae. Biol. Bull.* **40**, pp. 259–267.

—— (1923). A Study of the Chromosomes in Three Species of Pseudococcus. *Arch. f. Zellf.* **17**, pp. 45–62.

—— (1925). The Cytology of Pseudo-Sexual Eggs in a Species of *Daphnia. Zeit. indukt. Abst. Vererb.* **40**, pp. 1–27.

—— AND STURTEVANT, A. H. (1923). A Note on the Theory of Sex-Determination. *Amer. Nat.* **57**, pp. 379–381.

SCHWEITZER, J. (1923). Polyploidie und Geschlechtsverhältniss bei *Splachnum sphericum* Schwartz. *Flora*, **116**, pp. 1–72.

SEILER, J. (1914). Das Verhalten der Geschlechtschromosomen bei Lepidopteren. *Arch. f. Zellf.* **13**, pp. 159–269.

—— (1917). Geschlechtschromosomenuntersuchungen an Psychiden. *Zeit. indukt. Abst.* **18**, pp. 81–92.

SEREBROVSKY, A. S. (1922). Crossing-over involving Three Sex-linked Genes in Chickens. *Amer. Nat.* **56**, pp. 571–572.

—— (1925). Somatic Segregation in Domestic Fowl. *Jour. Genet.* **16**, pp. 33–42.

SHULL, A. F. (1910). Studies on the Life Cycle of *Hydatina senta*. I. Artificial Control of the Transition from the Parthenogenetic to the Sexual Method of Reproduction. *Jour. Exp. Zool.* **8**, pp. 311–334.

—— (1911). II. The Role of Temperature, of the Chemical Composition of the Medium, and of Internal Factors upon the Ratio of Parthenogenetic to Sexual Forms. *Ibid.* **12**, pp. 283–315.

—— (1912). Studies on the Life Cycle of *Hydatina senta*. III. Internal Factors influencing the Proportion of Male Producers. *Ibid.* **12**, pp. 283–315.

—— (1913). Nutrition and Sex Determination in Rotifers. *Sci.* **38**, pp. 786–788.

—— (1915). Periodicity of Male Production in *Hydatina senta. Biol. Bull.* **28**, pp. 187–197.

SIEBOLD, V. C. TH. (1864). Ueber Zwitterbienen. *Zeit. wiss. Zool.* **14**, pp. 73–80.

STERN, C. (1926). Vererbung im Y-Chromosom von *Drosophila melanogaster. Biol. Zentralb.* **46**, pp. 344–348.

STEVENS, N. M. (1905–6). Studies in Spermatogenesis. I, II. *Carn. Inst. Publ.* No. 36, pp. 1–74.

MECHANISM OF SEX-DETERMINATION   63

aSTEVENS, N. M. (1908). A Study of Germ Cells of Certain Diptera with reference to the Heterochromosomes and the Phenomena of Synapsis. *Jour. Exp. Zool.* 5, pp. 359–374.

—— (1909). Further Studies on the Chromosomes of the Coleoptera. *Ibid.* 6, pp. 115–124.

—— (1910). An Uneven Pair of Heterochromosomes in *Forficula. Ibid.* 8, pp. 227–242.

—— (1911). Heterochromosomes in the Guinea-Pig. *Biol. Bull.* 21, pp. 155–167.

STORCH, O. (1924). Die Eizellen der heterogonen Rädertieren. *Zool. Jahrb.* (Anatom.), 45, pp. 309–404.

STURTEVANT, A. H. (1920). Intersexes in *Drosophila simulans. Sci.* 51, pp. 379–380.

—— (1920). The Vermilion Gene and Gynandromorphism. *Proc. Soc. Exp. Biol. Med.* 17, pp. 70–71.

—— (1921). Genetic Studies on *D. simulans.* II and III. *Genet.* 6, pp. 43–64, 179–207.

TOYAMA, K. (1906). Studies on the Hybridology of Insects. I. On Some Silkworm Crosses with special reference to Mendel's Law of Heredity. *Bull. Coll. Agric. Tokyo,* 7, pp. 262–393.

WALLACE, L. B. (1905). The Spermatogenesis of the Spider. *Biol. Bull.* 8, pp. 169–184.

—— (1909). The Spermatogenesis of *Agalena naevia. Ibid.* 17, pp. 120–160.

WALTON, A. C. (1916). *Ascaris canis* and *A. felis.* A Taxonomic and Cytological Comparison. *Biol. Bull.* 31, pp. 364–372.

WASSILIEF, A. (1907). Die Spermatogenese von *Blatta germanica. Arch. mikr. Anat.* 70, pp. 1–42.

WEBER, M. (1890). Ueber einen Fall von Hermaphroditismus bei *Fringilla coelebs. Zool. Anz.* 13, pp. 508–512.

WENKE, K. (1906). Anatomie eines *Argynninia paphia* Zwitters. *Zeit. wiss. Zool.* 84, pp. 95–138.

WETTSTEIN, v. F. (1924). Morphologie und Physiologie des Formwechsels der Moose auf genetischer Grundlage. *Zeit. indukt. Abst.* 33, pp. 1–236.

WHITNEY, D. D. (1907). Determination of Sex in *Hydatina senta. Jour. Exp. Zool.* 5, pp. 1–26.

—— (1910). The Influence of External Conditions upon the Life Cycle of *Hydatina senta. Sci.* 32, pp. 345–349.

—— (1914). The Influence of Food in Controlling Sex in *Hydatina senta. Jour. Exp. Zool.* 17, pp. 545–558.

—— (1914). The Production of Males and Females controlled by Food Conditions in *Hydatina senta. Sci.* 39, pp. 832–833.

—— (1915). The Production of Males and Females controlled by Food Conditions in the English *Hydatina senta. Biol. Bull.* 29, pp. 41–45.

—— (1916). The Control of Sex by Food in Five Species of Rotifers. *Jour. Exp. Zool.* 20, pp. 263–296.

64 MECHANISM OF SEX-DETERMINATION

WHITNEY, D. D. (1924). The Chromosome Cycle in the Rotifer *Asplanchna intermedia*. *Anat. Rec.* 29, p. 107.
WILSON, E. B. (1905). Studies on Chromosomes. I. *Jour. Exp. Zool.* 2, pp. 371–406.
—— (1906). III. *Ibid.* pp. 1–40.
—— (1907). The Case of *Anasa tristis*. *Sci.* 25, pp. 191–193.
—— (1909). Studies on Chromosomes. IV and V. *Jour. Exp. Zool.* 6, pp. 69–100, 147–206.
—— (1911). The Sex Chromosomes. *Arch. d'anat. microsc.* 77, pp. 249–271.
—— (1913). A Chromatid Body simulating an Accessory Chromosome in *Pentatoma*. *Biol. Bull.* 24, pp. 392–410.
—— (1925). The Cell in Development and Inheritance 3rd edit. New York.
WINGE, O. (1922). A Peculiar Mode of Inheritance and its Cytological Explanation. *Jour. Genet.* 12, pp. 137–144.
—— (1923). Crossing-over between the X- and the Y-chromosomes in *Lebistes*. *Ibid.* 13, pp. 201–207.
WINIWARTER, V. H. (1912). Études sur la spermatogénèse humaine. II. Hétérochromosomes et mitoses de l'épithelium séminal. *Arch. d. biol.* 27, pp. 128–190.
—— (1914). L'hétérochromosome chez le chat. *Acad. Roy. Belg.*
—— AND SAINMONT, G. (1909). Nouvelles recherches sur l'ovogénèse et l'organogénèse de l'ovaire des mammifères. VI. Ovogénèse de la zone corticale primitive. *Arch. d. biol.* 24, pp. 165–267.
WODSEDALEK, J. E. (1913). Accessory Chromosomes in the Pig. *Sci.* 38, pp. 30–31.
—— (1913). Spermatogenesis in the Pig. *Biol. Bull.* 25, pp. 8–46.
—— (1914). Spermatogenesis in the Horse. *Ibid.* 27, pp. 295–304.
—— (1920). Studies on the Cells of Cattle, with special reference to Spermatogenesis and Sex-Determination. *Ibid.* 38, pp. 290–316.
WOLTERECK, R. (1911). Ueber Veränderungen der Sexualität bei Daphniden. *Inter. Rev. ges. Hydrob. u. Hydrogr.* 4, pp. 91–128.
WOODRUFF, L. L. AND ERDMANN, R. (1914). A Normal Periodic Reorganisation Process without Cell Fusion in Paramecium. *Jour. Exp. Zool.* 17, pp. 425–518.
YOCUM, H. B. (1915–17). Some Phases of Spermatogenesis in the Mouse. *Univ. Calif. Publ.* 16, pp. 371–380.
ZAVADOVSKY, M. (1926). The Mechanism of Differentiation of Sex Characters. *Trans. Lab. Exp. Biol., Moscow*, 2, pp. 29–83.
ZELENY, C. AND FAUST, E. C. (1915). Size Dimorphism in the Spermatozoa from Single Testes. *Jour. Exp. Zool.* 18, pp. 187–240.
—— AND SENAY, C. T. (1915). Variation in Head Length of Spermatozoa in Seven Additional Species of Insects. *Ibid.* 19, pp. 505–514.

Chapter III

THE PHYSIOLOGY OF SEXUAL DIFFERENTIATION

A "genotypic" male has the genetic constitution represented by the formula XY in the *Drosophila* type, XX in the *Abraxas* group; a "genotypic" female, XX, or XY respectively. The individual at first is but a zygote and much must happen before it can become a functional male or female. The complicated processes of sex-differentiation, during which the sex-organisation of the individual assumes one or other type, male or female, must be pursued before the one sex can be distinguished from the other by differences in anatomical structure, in physiological functioning and in psychological characterisation; before the genotypic male becomes the phenotypic male, the genotypic female the phenotypic female.

In the vertebrates the initial sexuality, which is but the direct expression of genetic action, is strongly reinforced when the components of the endocrine system become differentiated. The results of experimental gonadectomy and of the implantation of gonadic tissues have clearly demonstrated the intimate relation between the type and degree of sexual differentiation and the functional activity and histological organisation of the gonads (Steinach, 1913; Sand, 1921, 1923; Moore, 1921; Lipschütz, 1924). They have shown conclusively that for the complete differentiation and maintenance of the sexual phenotype active functional gonadic tissue is essential; that a male possessing testes, because the embryonic gonads assumed the testicular organisation for the reason that they were XY tissues and developed in the initial internal environment of maleness, becomes a phenotypic male because the testes exhibited a certain specific kind of physiological activity, elaborated a specific male "sex-hormone," which greatly reinforced the initial internal environment of maleness, and exerted, during the critical period of their development, a directing stimulus that evoked a ready response on the part of the remaining structures of the sex-equipment so that they pursued their differentiation according to the male type of sexual architecture. They have shown that an individual becomes a phenotypic female because it possessed functionally active ovaries at the time when the rest of the structures of the sex-equipment underwent differentiation. When it can be shown how it is that she comes to possess ovaries, there will remain no difficulty in showing how it is that she becomes equipped as an efficiently functional female.

It is not to be denied that for the efficient differentiation and maintenance of the sexual equipment the activity of other members of the endocrine chain is essential (Blair Bell, 1920; Bullock and Sequira, 1905; Glynn, 1911–12; Bénoit, 1924; et alia), but for the present it is reasonable to assume that the rôle of such glands as the adrenal, the pituitary, the thyroid, in sexual differentiation, is relatively unimportant when compared to that of the differentiated gonad. It is true that disfunctioning of any member of the endocrine system may be reflected clearly in a malfunctioning of the gonad or in the establishment of an imperfect sexual phenotype, but there can be no great satisfaction in following this trail until it has been better blazed.

Reviewing that which has been written above, the sexual characterisation in the higher vertebrates can be summarised, for purposes of discussion, as is shown on the opposite page. It is to be noted that in the case of the bird the female is the digametic sex (XY) and the male monogametic (XX).

Reasons for holding that the gonads in certain vertebrate groups, e.g. birds, do not function as endocrine glands, elaborating specific sex-hormones, will be considered later.

In the case of insects (Oudemans, 1899; Kellogg, 1904; Prell, 1915; Kopeć, 1912, 1922; Meisenheimer, 1909; Hegner, 1914) it has been established by experimental gonadectomy and the implantation of gonadic tissues that no part of the control of sexual differentiation is, in these forms, relegated to the gonad. All the sexual characters of the insect are secondary genotypic and owe nothing in their differentiation to the physiological activity of the differentiated gonad. The differentiation of a cell is pursued under the control of its own genotype.

Sex-dimorphic
characters

(1) *Primary genotypic sexual characters:*   *Male*   *Female*

    The sexual genotype—symbolised as—   XO or XY   or   XX
    in action establishes
    an internal environment—symbolised as— $1X : 2A$ or $1X : 1A$
    in which ontogeny proceeds and the
    sexual phenotype is assumed. This
    consists of

(2) *Secondary genotypic sexual characters:*
    These include: (A) Such sexual differences as are not
    secondary gonadic (many of them
    are epigamic, *i.e.* concerned only
    indirectly with, but contributory
    to, fertilisation, playing their part
    in courtship and the care of the
    (B) *Primary gonadic characters:*   young)

The following are bracketed together as "The gamic characters concerned directly with fertilisation":

    The *gonads* become differentiated   Testes   or   Ovaries
    The "sex-hormone" is liberated   Male sex-hor-   or   Female sex-
    and the   mone   hormone

(3) *Secondary gonadic characters* become
    differentiated:
    (a) The accessory sexual apparatus: Wolffian duct   Müllerian duct
    Müllerian ducts and Wolffian   derivatives   derivatives
    ducts   flourish   flourish

        Müllerian duct   Wolffian duct
        derivatives   derivatives
        cease their   cease their
        development   development

    (b) The external genitalia:
    Urogenital sinus becomes   Scrotum   or   Vulva
    Phallus becomes   Penis   or   Clitoris

    (c) Certain of the epigamic characters   *E.g.* regional distribution of hair,
    which for their expression depend   mammary glands, voice
    upon the stimulus of the differ-
    entiated gonad, and become ex-
    pressed at different times as they
    reach the stage of their develop-
    ment when they can respond

Finkler (1923) has carried out some extremely interesting experiments with various Orthoptera (*Hydrophilus, Dysticus, Notonecta, Dixippus, Tenebrio, Vanessa*). He transplanted male heads on to female bodies and *vice versa*, and observed the effect of this upon sexual behaviour. In the case of *Hydrophilus*, three operations were entirely successful, complete healing occurring in about seventeen days when all movements were properly co-ordinated and feeding and defaecation were normal. A female with a male head behaved sexually as a male towards females, and a male with a female head as a female, but was regarded by normal males as a male. In certain instances the body took on the coloration of the sex to which the transplanted head belonged. Blunck and Speyer (1924), repeating this work, failed to confirm Finkler's findings.

The clearest view of the almost virgin field of developmental physiology can be gained from a study of certain kinds of ontogenetic intersexuality.

If in an individual of a group in which bisexuality is customary, maleness and femaleness are exhibited coincidently or in succession, and to any degree, that individual is an intersexual form. It is an intersex either in time or in space. Intersexuality is the state or quality exhibited by an individual of a normally dioecious group in which both maleness and femaleness are to be distinguished in varying degree and/or at different times. Hermaphroditism is the expression of a sexuality normal to a particular group; intersexuality, transient or permanent, an expression of sexual abnormality in a particular group or individual. Hermaphroditism and intersexuality are the same expression of the self-same mechanism but whereas intersexuality is perfect or imperfect hermaphroditism in a non-hermaphroditic group or individual, hermaphroditism is perfect intersexuality in a group or individual in which such a state or quality is a customary feature of the normal life history.

From what is known of the mechanism of sex-determination and sexual differentiation, it can be anticipated that intersexuality in a normally dioecious group will follow upon (1) disharmony in the sexual genotype of the body as a whole or of local parts, and (2) profound alteration of the internal environment during critical stages of development.

A. INTERSEXUALITY DUE TO A DISHARMONY IN THE COMPOSITION OF THE GENOTYPE.

(1) *Serial Intersexuality in* Lymantria.

It has long been known (Standfuss, 1896) that when species or even geographical races of Lepidoptera are crossed sexual abnormalities commonly are found among the hybrid offspring. For example, if European specimens of the notorious forest pest, *Lymantria dispar*, the gipsy moth, are bred among themselves the offspring are unremarkable in every way. The same is true in the case of the Japanese variety, *Lymantria japonica*. But if a Japanese male is mated with a European female, normal male offspring and females that show a number of modifications in the direction of maleness, *i.e.* females that are intersexual, are produced. When such female intersexes were fecund, *i.e.* were not too abnormal, and were mated with their brothers, it was found that in the succeeding generation typical Mendelian segregation occurred, half the females being sexually normal, half being intersexual. The reciprocal cross, European male and Japanese female, produces normal males and females in the first cross-bred generation, but if these are then interbred there appear in F_2 a certain proportion of males that show modification in the direction of the female type of organisation, *i.e.* males that are intersexual.

Further investigation by Goldschmidt (1923) and his associates demonstrated that there were many different races of European, Japanese, and American gipsy moths that were quite distinct in their hereditary constitution in respect of this character intersexuality, in that the degree or grade of intersexuality was definite and typical for a particular mating. As the result of much experimental breeding work Goldschmidt was able to classify his strains as relatively "strong" or "weak." For example, a "strong" male mated to a "weak" female would give 50 per cent. normal males and 50 per cent. intersexual females. A "very strong" male mated to a "weak" female would give offspring all possessing the male type of sexual organisation. A mated to B gave a low grade of intersexuality, C × D a high grade, E × F a grade intermediate between these, and so on. If strong race A gave moderate inter-

sexuality with weak race P, whilst with race Q it gave strong intersexuality, and if strong race B gave moderate intersexuality with
Q, then it could be predicted that B with P would give only a
slight grade of intersexuality. Similar males from one culture
mated with females from different cultures gave intersexes that
could be arranged in a series according to the degree of their
abnormality: males from different cultures mated with females
from one and the same culture gave intersexes that differed from
these but which could also be arranged in a series. It was possible,
by calling on experience, to produce every stage from an almost
complete male to an almost complete female intersex at will by
making the appropriate mating. In fact, it was as possible to turn
the genotypic females into phenotypic males as to ensure the regular
production of normal males and females.

It was noticed that the condition of intersexuality did not affect
all the structures of the sexual organisation equally. Some were
normal, whilst others were intersexual in their characterisation,
and only purely quantitative characters such as the length of the
antennal pennae exhibited a condition intermediate between those
of the normal male and female. Further investigation demonstrated that the different structures could be arranged in a definite
series as regards degree of intersexuality in characterisation and
that this series was exactly the opposite to the order of the
embryonic differentiation of these structures. Those organs which
are the first to be developed and differentiated are the last to be
modified; those that appear last are the first to be changed. From
these considerations there arose the Time Law of Intersexuality.
An intersex is an individual that has developed as a male (or female)
up to a certain point in its life history and thereafter has continued
its development as a female (or male). The degree of intersexuality
is determined by the time during the critical period of sexual
differentiation at which this switch-over occurred.

The intersexual females start their development as females and
then at a certain point in their development change their mode of
differentiation and finish as males, and since the hard parts of an
insect are external and composed of chitin any of them that have
hardened before the switch-over remain unaltered by it. From
an examination of the parts which are sexually dimorphic, it is

possible to decide in the case of any particular individual exactly when the change-over took place. The intersex is a sex-mosaic in time. The female showing the least degree of intersexuality has only the feelers modified in the manner of the "feathered" male type. Those displaying a further degree of intersexuality will have the male colouring of the wings in addition. A further stage will consist in the addition of a male-type of copulatory organs. Next will be a male-type abdomen, and the final stage of all will be that in which the ovaries are replaced by testes, *i.e.* the genotypic female (according to chromosome content) will have become a phenotypic male, *i.e.* an XY individual has assumed the characterisation normal in one with the XX type of chromosome constitution. The fact that an individual primarily equipped with ovaries can become so transformed as to possess testes is of profound interest since it illustrates in the clearest possible manner the relation of the structure of the gamete to the organisation of the individual producing it. Genotypically the sperm will be identical with the ova previously produced but structurally and physiologically they will be profoundly dissimilar. The genotype of the zygote decides the organisation of the gonad and this decides the morphological and physiological properties of its manufactured product, the egg or the sperm. It is not the genotype that distinguishes egg from sperm; they differ because they are the products of dissimilar gonadic organisations and the difference takes the form of dissimilar structure and behaviour.

From these facts Goldschmidt deduced the following conclusions:

(1) Each sex possesses the potentialities of the other since either can become intersexual. (In other words, the embryonic tissues are ambivalent as regards their future sexual differentiation.)

(2) The type of sexual differentiation that the zygote will pursue is determined at the time of, and by the mechanism of, fertilisation. (In other words, sex is determined by the nature and interaction of the genotype established in the zygote by the conjugating gametes. If the genetic constituents of intersexuality are in the zygote, then the individual will inevitably become an intersexual form.)

(3) The normal determination of sex is bound up with the X: 2X mechanism. But as this does not prevent the occurrence of intersexuality and the transformation of one sex into the other, it cannot be the mere presence of these chromosomes or the factors contained within them that counts, but rather their quantitative effect. (This is in line with the conclusions that have emerged from the consideration of other forms of intersexuality discussed earlier in chapter II.)

(4) The mode of inheritance of this intersexuality shows that since the female in *Lymantria* is XY, and since her single X-chromosome is received from her father, one of the deciding factors in sex-determination is transmitted in the X-chromosome. (This is in line with previously formed conclusions.)

(5) Other factors concerned in sex-determination in *Lymantria* are purely maternal in inheritance, being resident in the Y-chromosome. A daughter receives her Y-chromosome from her mother. But since a male has no Y-chromosome, the factors in the Y-chromosome must have exerted their action on the unripe egg when this contained both X- and Y-chromosomes. (It will be remembered that the mature egg of a female that is digametic carries either an X or the Y. This is due to the fact that of this XY pair one member passes into the polar body whilst the other stays in the egg. The eggs will therefore be of two sorts, X-chromosome bearing and Y-chromosome bearing, the former on being fertilised by the X-chromosome-bearing sperm giving rise to an XX individual, a male, the latter being fertilised, to an XY zygote, a female.) If all eggs are to be alike in respect of the Y-borne genes, these must have acted and their products must have specifically affected the cytoplasm before the X and Y became disjoined. It is seen then that there is good reason for postulating that different genes come into play at different times. It is to be noted that the Y-chromosome of *Lymantria* is not as the Y of *Drosophila* but resembles the Y of fishes in that it carries genes affecting sex-determination. (Aida, 1921.)

(6) The fact that the genetically similar females give different results when mated with genetically dissimilar males shows that the sex-determining genes in the X-chromosome differ quantitatively in the different races. The fact that genetically similar males

give different results when mated with genetically dissimilar females shows also that the sex-determining factors resident in the Y-chromosome can be different quantitatively.

It will be seen that if both X-borne and Y-borne sex-determining genes can so vary quantitatively among themselves, an infinite variety of different genetic combinations can be made, deliberately or by chance. In order to explain his results and his hypothesis Goldschmidt developed a somewhat intricate presentation. To the male-determining tendency of a particular race he assigns a positive value that is proportional to the strength of the male-determining genes. To the female-determining tendency he assigns another value also proportionate to the strength of the female-determining genes. He then assumes that when in an individual the male value is greater than the female value by a certain number of units the individual is a male, and that conversely when the female value is greater than the male by a certain number of units the individual is a female ($F - M = > 20$, a female; $MM - F = > 20$, a male). In a weak race, for example, F, the female-determining gene-complex, has a value of 80 units assigned to it and M has 60. Since the male-determining gene-complex M is borne on the X-chromosome and the female-determining gene-complex F is in every egg in the female, XY in constitution, F 80 exceeds M 60 by the epistatic minimum of 20 units, and the individual therefore becomes a female; in the XX males, on the other hand, $M = 60 + 60$ as against $F = 80$, so that there is an excess of 40 units in the male direction. In a strong race F is 100, M is 80, but the arithmetical relation between the values of F and M still govern the sexuality of the individual. (It should be stated that it does not seem impossible to restate the whole argument in terms of a numerical ratio as Bridges has done in the case of *Drosophila*. This algebraic interpretation is not really satisfactory.)

The conception of Goldschmidt can be illustrated by assigning arbitrary numerical symbols to the sex-determining factors. M = male-determining gene resident on the X-chromosome. M_1, M_2, M_3, M_4, M_5, etc., are male-determining genes of relatively different efficiencies; M_1 is a weak M, M_5 a relatively very strong M, and so on. Since a male has two X-chromosomes he carries two

PHYSIOLOGY OF SEXUAL DIFFERENTIATION

M's and these may be of different grades of efficiency, since they come from different sources. The following example may be cited.

A "weak" ♂	$M_2M_2F_3$	M_3mF_4	A "strong" ♀
Gametes	M_2	$M_3F_4 : mF_4$	Gametes
F_1	$M_2M_3F_4$	M_2mF_4	F_1
	♂ (M > F)	♀ (M < F)	
Gametes	$M_2 : M_3$	$M_2F_4 : mF_4$	Gametes
F_2 $M_2M_2F_4$	M_2mF_4	$M_3M_2F_4$	M_3mF_4 F_2
Intersexual ♂ (M = F)	♀ (M < F)	♂ (M > F)	♀ (M < F)

By the use of such symbols it is easy to illustrate the production of an F_1 generation consisting of 50 per cent. males and 50 per cent. females whose bodily characteristics are male:

P_1 "strong" ♂	$M_5M_5F_6$	M_2mF_3	"Weak" ♀
Gametes	M_5	$M_2F_3 : mF_3$	Gametes
F_1	$M_5M_2F_3$	M_5mF_3	F_1
	♂ (M > F)	♀ (M_5mF_3)	
		(but M > F, so all with	
		♂ organisation)	

Goldschmidt infers with reason that the sexual characterisation of any particular organ of the sex-equipment depends on whether one or the other type of sex-differentiating substance is effectively in excess at the time when the organ arises in development. He interprets the mosaic character of the intersex on the assumption that the amount of sex-differentiating substances produced in virtue of the presence of the corresponding sex-determining genes is not constant throughout life: that at one time the male-differentiating substance is in excess, at another the female. In the male of the moth M > F and the male-differentiating substance is effectively in excess until the period of development is complete. In the female M < F and the female-differentiating substance is effectively in excess during development. But if it should so happen that the sex-differentiating substances are produced at different rates, and if some genes possess the property of producing more sex-differentiating substance in a given time than others, then there exists the possibility that sex mosaics in time will be produced.

A male of a race in which the sex-determining genes work at a faster rate is crossed with a female of a race in which these genes work at a slower rate. The female-determining factor F is always inherited through the mother and in all the offspring there will be this factor F and the female-differentiating substance will be produced in all at the same rate. The male offspring, on the other hand, will receive one M from their mother, and the other, the quickly producing M, from their father, so that in a given time

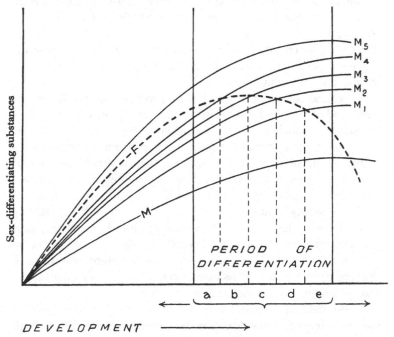

Fig. 21. The Time Law of Intersexuality. (*After* Goldschmidt.)

the male-differentiating substance will be effectively in excess during development. The female offspring will have the M from the father, and as a consequence the amount of male-differentiating substance will increase relatively to the amount of the female-differentiating substance, overtake it, and finally supplant it, and from this point onwards any sex-characters which still have to develop will be of the male type. The individual will be a female intersex. It is not the absolute but the relative rates of production of male- and female-differentiating substances that control the modelling of the sex-equipment.

There are two sorts of sex-determining genes, male-determining, borne on the X, and female-determining borne on the Y. The male genetically is MMF and the female MmF, since the action of the Y-borne genes is prezygotic. The sex-determining substances elaborated by these genes are known as male-differentiating and female-differentiating substances. These are elaborated at different rates by different male-determining and female-determining genes respectively; the introduction of more quickly and of more slowly elaborating genes into a genotype can result in a disturbance of the previous quantitative relation of the two sorts and so yield the intersexual condition.

In general the point at which the curves for the production of male- and of female-differentiating substances intersect lies beyond the stage at which differentiation occurs. But if the growth rate could be changed so that the point of intersection occurred during or before the period of differentiation, then it should be possible to produce a partial or complete reversal of sex and to produce individuals genetically of one sex (XX or XY, that is), but somatically, phenotypically, of the other. Goldschmidt by rearing *Lymantria* at very low temperatures produced intersexes, thus verifying this inference. Sex-reversal in these cases is due to genetic causes—the fertilised egg contains inevitably within itself the seed of its eventual transformation in the form of a quantitative disharmony of the sex-determining factors.

In the case of *Lymantria* the situation can be summarised as follows. Genetic intersexuality is the condition in which as a result of genetic causes the differentiation of the sex-organisation of a genotypic male or of a genotypic female, having been pursued up to a point in a manner appropriate to that sex, is switched over, so that after this point it follows the plan appropriate to the opposite sex. An intersex is an XX (or XY) individual which, as a result of an abnormal sexual differentiation comes to possess more or less completely the sex-equipment appropriate to the alternative XY (or XX) individual. Not every part of the body is involved in this condition; only the structures of the sex-equipment are concerned and of these only those are affected the differentiation of which is not complete at the time of the switch-over; the earlier in development the switch-over occurs, the more structures will be affected and the greater will be the degree of the intersexual condition, and *vice versa*.

Intersexuality results when the male and female sex-differentiating genes are quantitatively incorrectly in harmony, one set with the other. It does not occur when, as in the case of the genotypic female which attains a typical female organisation, the female-differentiating reactions are predominant throughout the period of differentiation. Intersexuality occurs when during the course of this phase of differentiation the male-differentiating reactions overtake and supplant the female-differentiating reactions and so control the remainder of the process, or when the female-differentiating reactions replace the previously predominating male reactions.

Under ordinary circumstances the female sex-differentiating substances in the male or *vice versa* are produced so slowly that an effective quantity is not present until the critical phase of differentiation has closed. If, however, their production were speeded up through the presence and action of a more rapidly producing female gene-complex, the time point at which these female-differentiating substances attained effective supremacy would fall within the critical period of differentiation and would then be the switch-over point. Each fertilised egg normally possesses male and female-determining genes which elaborate enzyme-like and specific sex-differentiating substances. In species in which the female has the constitution symbolised as XY, the female sex-differentiating substances are purely maternal in their mode of transmission so that every egg produced by one and the same female is identical in respect of this. The male-determining gene-complex is X-borne and therefore is present in half the eggs and in all the sperm. The rate at which the two sex-differentiating substances are elaborated (or the time at which they come into action) is a fixed heritable character of a race.

The case of *Lymantria* is of particular interest because it provides strong evidence that there are genes on the Y-chromosome and that different genes come into action at different times during development. It is seen that in *Lymantria* the Y-borne genes are not concerned with the development of morphological characters directly but with the establishment of a certain metabolic level. In the case of *Drosophila* it was long thought that the Y-chromosome was not concerned in the development of morphological characters but that its presence was necessary if the male was to be fecund. It is possible that most of the genes on the Y-chromosome of *Drosophila*, like the Y-borne genes of *Lymantria*, are concerned with the establishment of physiological characters and exert their action before organogeny commences.

It is important to note that though the phenomena of intersexuality in *Lymantria* are commonly interpreted as the results of genetic difference in the rate of production of sex-differentiating substances, it is equally satisfactory to explain them as the sequelae of genetic differences in the time at which different gene-complexes concerned in sex-determination come into action during development. M_1 and M_5 would then be equally potent, but M_5 would come into action earlier.

The researches of Pflüger (1882), Hertwig (1912), Schmidt-Marcel (1908), Kushakevitch (1910) and Witschi (1914, 1915) have revealed the fact that in the frog there are two forms of development of the reproductive organs. In one the differentiation of the gonad is either that which leads directly to the organisation of an ovary or that which directly yields a testis. In the other there is in the beginning an organisation that is female in its characterisation but sooner or later 50 per cent. of these cases become transformed into testes. It has been established that this tendency to early and direct differentiation on the one hand, and to late and indirect differentiation on the other are truly heritable characters and that the hereditary factors for these characters can be transmitted by either father or mother; breeding experiments yielded results closely resembling those obtained in the mating of individuals of different "efficiencies" in *Lymantria*, and there is good reason for interpreting these temporary intersexuals in terms of Goldschmidt's *Lymantria* hypothesis. Swingle (1925) very ably upholds the view that the early gonad of the male of indirect and late differentiation is not ovarian and that such a male is not an intersex. He argues that this progonad is sexually neutral in consequence of a balance between male and female determining tendencies in the zygote, and the fate of the progonad is decided when the definitive sex-cords appear, for then the progonad becomes transformed into a testis. The definitive testis cannot develop in the absence of sex-cords and the appearance of these decides the course of development. The time of the appearance of these peculiar male structures, the sex-cords, is probably determined by genetic factors and so different races and their hybrids differ in the time at which progonad becomes testis. According to Swingle "Pflüger's hermaphrodites" are but males in which the appearance of sex-cords in the progonad is greatly retarded, and many recorded cases of sex-reversal in the frog and toad are but instances of such retardation.

It is of considerable interest to note that Witschi secured an adult hermaphrodite the eggs and sperm of which were used. The eggs of this hermaphrodite fertilised with sperm from an individual of a race in which early and direct differentiation was the rule yielded males and females: sperm of the hermaphrodite fertilising eggs of a female of a race in which early and direct differentiation was the rule gave only females. These facts point to the conclusion that this hermaphroditic frog was XX in sex-chromosome constitution and that all its gametes, eggs and sperm alike, contained a single X. The X sperm fertilising the X eggs of the monogametic female yielded only XX individuals: the X eggs being fertilised by both X and Y sperm gave both males and females. It is to be noted that here again is an instance of egg and sperm being genotypically equivalent, of ovary and testis being possessed of the same genotypic constitution. Manifestly though an individual that has the sex-chromosome constitution XY develops testes, these develop not because they are XY but because they developed in a particular kind of internal environment that the interaction of the genes in an initial XY genotype established.

Hertwig and Witschi were able to show that temperature has a definite influence on the type of gonadic differentiation in frogs. They found that certain forms cultivated at 21° C. showed early and direct gonadic differentiation, whereas when cultivated at 27° C. some of the females changed after the metamorphosis into males. In another race cultivated at 27° C. all these females became transformed into males. Evidences of this transition in sexual differentiation can be obtained from other sources. At the anterior end of the testis of the toad is Bidder's organ, a rudimentary ovary. It is found in both sexes whilst the individuals are immature but in the case of the female it degenerates. Harms (1923) was able to show that Bidder's organ was a rudimentary ovary and that it possessed an endocrine function. All the males of *Perla marginata* possess a well-developed but non-functional ovary anterior to the testis (Schönemund, 1912). In *Phyllodromia germanica* Heymons (1890) has demonstrated that the anterior part of the gonad of the male is differentiated into ovarian tissue: in *Orchestia* Nebeski (1881) has shown that a part of testis always contains eggs, whilst in the crab, *Gebia major*, Ischikawa (1891) has shown that the anterior part of the gonad of the male is testicular whereas the posterior portion is ovarian though incapable of functioning. In *Myxine* Schreiner (1904) has shown that whilst no real hermaphroditism occurs, the gonad is mixed, though only the testicular or the ovarian portion is functional. The work of Grassi (1919) on the eel suggests that similar conditions obtain in this form. These and many other cases can be interpreted by an appeal to the hypothesis of Goldschmidt.

It has been tacitly assumed that an oocyte or a spermatocyte possesses its peculiar cytological features in virtue of its own nuclear sex-chromosome constitution. But such a dogma is becoming more and more untenable, since it fails to account for a number of well-authenticated facts and is unsupported by critical evidence. Actually, an oocyte is a germ-cell in the meiotic phase characterised by certain features, of which the principal are the intercalation of a diffuse post-diplotene stage, with the active enlargement and proliferation of the plasmosome, and growth of the cytoplasm correlative to it. These features are undoubtedly not entirely restricted to oocytes, for in the spermatocytes of such forms as *Saccocirrus* and of some Hemiptera a very transient diffuse stage is seen; whilst in the protestis of *Bufo* and in the functional male gonad of some chilopods, all the characteristic features of the oocytes are to be found.

A more reasonable view would seem to be that the cytological features of the oocytes or of the spermatocytes depend, just as does the character of the somatic structures of the rest of the sex-equipment of the individual (cf. the differentiation of the sex-mosaics of *Lymantria*), on the balance of conflicting physiological factors determined by the sex-chromosome constitution of the body as a whole, or, in the case of mammals, locally through the action of the interstitial cells of the gonad.

On this assumption Goldschmidt's conception of a timing mechanism puts the discussion as to whether ovum-like bodies in a testis are oocytes or not on an entirely new footing and shows that such questions as the homology of Bidder's organ to an ovary or the transformation of a characteristically female definitive gonad to one of the male type does not in the least conflict with the view that genetical factors play an important rôle in sex-differentiation. If it is agreed that the essential difference between the male and the female lies in the timing mechanism which decides whether, whilst a given organ is developing, the male or the female-determining reactions are predominant, and if it is agreed that at some point in the development preceding or following the stage of differentiation the female-determining reactions are predominant in a "genotypic" male and *vice versa*, then such questions as the above are resolved into a mere verbal quibble, and the efficiency of environic agencies to co-operate with the genetical factors offers no difficulty. The provisional hypothesis outlined by Goldschmidt to account for his *Lymantria* intersexes then brings into one coherent scheme undoubted sex-transformation as seen in *Crepidula*, *Sacculina*, and *Bonellia*, the occurrence of oviform cells in the protestis of Anura (or among myriopods as steps in normal spermatogenesis), and the cases of sex-reversal met with in birds.

Fig. 22 shows how the significance of Bidder's organ, for example, can readily be interpreted on Goldschmidt's hypothesis. A represents the progonad; B the definitive testis. It will be remembered that the development of the progonad is actually later in the forms with a protracted larval phase. During the earlier part of the development of the individual the female-determining reactions are predominant, the female-differentiating substances are effectively in excess, and any organ that is developed during this period and is responsive to the action of their stimulus will pursue its development and differentiation under their direction; later, in the case of the determined male, the male-determining substances are effectively in excess, and all differentiation of the structures of the sex-organisation will thenceforward be according to the male plan. Thus it will happen that in the case of the anuran with a short larval phase, the progonad will present the characteristic features of Bidder's organ, whereas in the case of the anuran with a protracted larval life, in which the progonad is developed later, there will be no Bidder's organ and no ovum-like bodies in the testis.

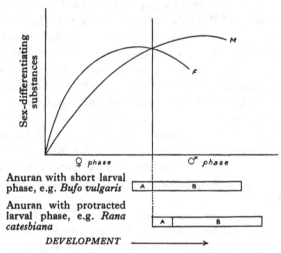

Fig. 22. Suggested application of Goldschmidt's hypothesis to the case of the *Anura*. A = progonad; B = definitive testis.

Other somewhat similar instances of intersexuality that must await interpretation until it has been shown whether they fall into the *Drosophila* or *Lymantria* categories are:

(1) The Biston species hybrids described by Harrison and Doncaster (1914) and by Harrison (1919).

(2) The sex intergrades of *Simocephalus vetulus* described by Kuttner (1909), by Banta (1916) and also by de la Vaulx (1921).

(3) The intersexual forms of *Gammarus chevreuxi* described by Sexton and Huxley (1921).

(4) The intersexual lice described by Keilin and Nuttall (1919).

(5) The intersexual Nematodes described by Steiner (1923).

(2) *Intersexuality in Vertebrates.*

In the mammal it is generally accepted that the reproductive glands function as ductless glands providing internal secretions which direct the development of the remaining structures of the sex-equipment. In them the sex-determining gene mechanism results in the production of the sex-differentiating substances, possibly enzyme-like in constitution, (in the establishment of a certain metabolic level), and these direct the differentiation of the embryonic gonad into testis or into ovary. The gonad becomes differentiated and in it is developed the gametogenic tissue which provides the gametes, and interstitial tissue which provides the male or the female sex-hormone, specifically different, that directs the development of the appropriate accessory sexual apparatus and secondary gonadic characters. Every individual, XX and XY alike, at the beginning of the development of its sexual organisation possesses paired undifferentiated gonads, paired Müllerian and Wolffian ducts, and structures which are concerned in the later development of the secondary gonadic characters of both sexes. The first stage in sex-differentiation is the formation of ovaries or of testes from undifferentiated gonads. Next, under the direction of the hormone produced by testis or ovary, either the Müllerian or the Wolffian ducts continue their development to become the functional accessory sexual apparatus, and later, about the time of sexual maturity, under the influence of the sex-hormone, other appropriate secondary gonadic characters are displayed. If, as appears probable, the sex-hormone is chemically identical throughout the life history of the individual, then the assumption of the different sexual characters at different stages of the life cycle must imply that more than the sex-hormone is involved, or that interaction between gonad and the other endocrine glands such as pituitary and adrenal may be necessary, or that the different tissues respond only at a certain stage of growth which is not attained synchronously by all. The genotypic female, in the absence of disturbing agencies, develops ovaries: she develops these because she is a genotypic female, and is not a genotypic female because she has ovaries. The phenotypic female, on the other hand, is a phenotypic female because she has developed ovaries.

(a) *Intersexuality due to Abnormality in the Time of the Differentiation of the Gonads.* (*Pseudo-Intersexuality.*)

Abnormality of the reproductive system taking the form of an intimate mixture of male and female structures belonging to the accessory sexual apparatus is not uncommon among the domesticated mammals, and many cases have also been recorded in the human subject. The typical history in these cases is that an individual, regarded as a female during the earlier part of its life, later assumes many of the characters of the male. This peculiar type of abnormality is particularly common in the goat and pig. In the goat cases are known in which an individual which actually won prizes when shown as an immature female, from the time of sexual maturity became more and more like the male: its beard grew, its head became male-like, and about it there hung the pungent smell so characteristic of the male. In its behaviour it resembled the "rig," a male with maldescended testes, but its external genitalia retained the form of a vulva-like aperture with an over-large clitoris. When the internal genitalia from such a case are examined there are found paired gonads lying in the situation of ovaries or somewhere along the track of the migrating testes, which on section show the structure typical of the maldescended testis; and an accessory sexual apparatus composed of more or less well-defined epididymes, vasa deferentia, seminal vesicles, prostate, Cowper's glands, uterus, and vagina, the latter in some cases having failed to establish communication with the vulva.

The literature contains many references to this type of abnormality in the human, goat, and pig, but these were treated as isolated examples and no successful effort was made to give to them a common interpretation until Crew (1923) examined a considerable number of cases in the goat, pig, horse, cattle, sheep, and camel. In a few of the cases examined, the external genitalia had the form of an unremarkable vulva and clitoris; in others the erectile organ, though female in form, was unusually large and prominent; in others it was distinctly peniform yet imperfectly canaliculised. In a few instances there was a solid conical elevation on the abdominal wall where in the normal male the penis ends.

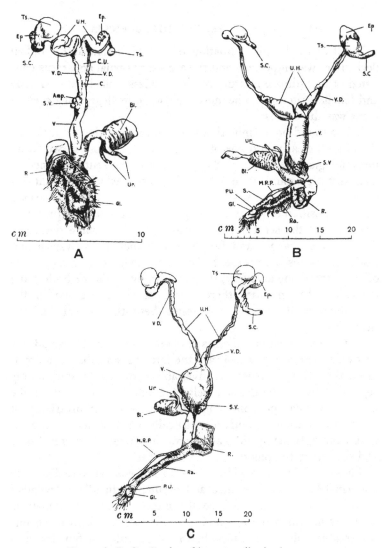

Fig. 23 A, B, C. Grades of intersexuality in the goat.

Amp.	Ampulla of vas deferens.	S.	Flexures of corpus cavernosum
Bl.	Urinary bladder.		penis.
C.U.	Body of uterus.	S.C.	Spermatic cord.
C.	Clitoris.	S.V.	Seminal vesicle.
Ep.	Epididymis.	Ts.	Testis.
Gl.	Glans penis.	U.H.	Uterine horns.
M.R.P.	Retractor penis.	Ur.	Ureters.
P.U.	Processus urethrae.	V.	Vagina.
R.	Rectum.	V.D.	Vas deferens.
Ra.	Median raphe of perineum.		

In all cases the differentiation of the Müllerian and Wolffian derivatives was imperfect, and the accessory sexual apparatus consisted of an intimate mixture of more or less well-developed male and female structures. The most variable in its degree of development was the uterus.

It was found on examination that the cases of this condition in the pig fell clearly into one of two classes: (1) those in which no morphological evidence of the previous or present existence of ovarian tissue could be found in the gonads, which were entirely composed of testicular tissues with a histological structure varying with the position of the testis along the line between the primitive position and the scrotum, but always exhibiting some degree of degenerative change; and (2) those in which both ovarian and testicular tissues were present, the gonads being one an ovary, the other a testis; one an ovary, the other an ovotestis; or both being ovotestes. In one case there were paired ovaries within the abdominal cavity and paired testes beneath the skin of the perineum.

It was noted that in the cases of class 2 in which one gonad was an ovary, this gonad was always the left one, and that in an ovotestis the ovarian tissue was invariably cephalad to and sharply separated by a well-defined belt of connective tissue from the caudal testicular portion. The ovarian tissue was invariably of apparently normal structure histologically, whereas in the testicular tissue the spermatogenic was always degenerate to some degree and the interstitial plentiful.

Those individuals in which the external genitalia were distinctly abnormal had been identified as "hermaphrodites" or "wilgils" by the breeders, and had been slaughtered while still immature. The few in which the external genitalia had the form of an unremarkable vulva and clitoris had been regarded as females until about the time of sexual maturity, when it was noted that instead of assuming the sexual characterisation of the adult female they had begun to exhibit the secondary gonadic characters of the male type, and that the clitoris had begun to increase in size. The sexual behaviour of such of these individuals as proved later to belong to class 1 was as that of the "rig," a male with maldescended testes. The behaviour of such as belonged to class 2 was imperfectly male. They were slaughtered because of their curious behaviour and because they failed to breed.

The fact that in one class of these cases of reproductive abnormality there is no ovarian tissue and no suggestion that there ever had been any, whereas in the other class there is ovarian tissue of apparently normal structure, points to the conclusion that the two classes are different in their nature. If ovarian tissue had ever been present in the case of class 1, it should have persisted, since in those cases in which ovarian tissue is present it is invariably of normal structure. It is proposed to regard the two classes as being distinct and to treat them separately.

During ontogeny there is a period in which the differentiation of the sexual organisation is timed to take place. At the beginning of this period, which follows a preliminary phase of growth and organ formation, the reproductive system consists of (1) paired gonads of indifferent histological structure; (2) a rudimentary accessory sexual apparatus composed of Müllerian and Wolffian ducts; (3) external genitalia represented by the growing urogenital sinus and genital tubercle. From this initial type of reproductive architecture, possessed in common by all individuals, genotypic male and female alike, one or other type of differentiated sexual organisation, male or female, is attained. The indifferent gonads become testes or else they become ovaries; if they become testes then the Wolffian ducts continue their development to become the functional deferent ducts of the testes, while the further development of the Müllerian ducts ceases, and the external genitalia become scrotum and penis. If the indifferent gonads become ovaries the Müllerian ducts continue their development to become the functional uterus with cornua and Fallopian tubes and vagina, while the development of the Wolffian ducts ceases, and the external genitalia assume the form of vulva and clitoris. Sexual differentiation is alternative and the end-product is an organisation appropriate to the functional female or else to the functional male.

Between the differentiation of the various structures of the sex-equipment there is a time relation. The first structure to begin its differentiation is the gonad, the next, it is assumed, the external genitalia, and finally the structures of the accessory sexual apparatus. The results of gonadectomy and gonad implantation have shown perfectly clearly that for the appropriate differentiation of the rest of the sex-equipment the differentiated gonad is necessary. In the presence of functional testicular tissue the sexual organisation appropriate to the functional male is assumed; in the presence of functional ovarian tissue that appropriate to the functional female. Such differentiation is pursued under the control of specific male and female sex-hormones elaborated by the testis and ovary respectively. Since the different structures of the sex-equipment respond to the physiological stimulus of the sex-hormones at different times during ontogeny it can be assumed that the threshold of response to this stimulus differs in different cases, and that before this threshold of response is reached a certain degree of undifferentiated growth is required. If the effect of the functioning of the differentiated testis is to inhibit the further development of the Müllerian ducts and their derivatives and to encourage the fuller development of the derivatives of the Wolffian ducts and to model the growing urogenital sinus and genital tubercle into scrotum and penis, then it is necessary only to explain the differentiation of the embryonic gonad into testicular tissues in order to explain the complete assumption of a male type sexual organisation by the individual.

This can be done if it is assumed that the gonad in its indifferent stage is ambivalent as regards its future mode of differentiation (though not completely so since its tissues are chromosomally either male or female, XO or XX), and that this differentiation is pursued under the direction of male-differentiating and female-differentiating substances elaborated by the male-determining and female-determining factors respectively. In the genotypic male (XO in sex-chromosome constitution) it is the rule for the male-differentiating substances to be effectively in excess over the female-differentiating substances during that period of development when the differentiation of the gonad is timed to take place, whereas in the genotypic female the female-differentiating substances are effectively in excess during this period.

These suggestions can be illustrated graphically as in Fig. 24 A and B.

The interpretations of the conditions found in class 1 can now be attempted. In these cases there were paired maldescended

testes, more or less well-developed derivatives of both Müllerian
and Wolffian ducts, external genitalia ranging from those of the
apparently normal female to those of the grossly imperfect male.
The cases could be readily arranged in a series according to the
degree of imperfection of the external genitalia and of the degree
of development of the Müllerian duct derivatives. This fact suggests that they are one and all but grades of the same condition
and that between them there exists a time relation. It must be
stated clearly that there was no evidence which suggested that the
abnormal individuals were free-martins[1] (see page 99), cases of

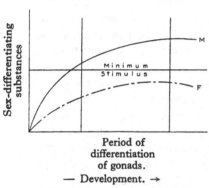

Period of
differentiation
of gonads.
— Development. →
In the normal male, the male-differentiating reactions are effectively in
excess during the whole period of
gonadic differentiation, and so the
gonads assume the structure of testes.
Fig. 24 A.

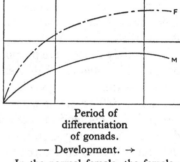

Period of
differentiation
of gonads.
— Development. →
In the normal female, the female-
differentiating reactions are effectively
in excess during the whole period of
gonadic differentiation, and so the
gonads assume the structure of ovaries.
Fig. 24 B.

sex-reversal in a genetically determined female (see chapter v), or
cases of true "glandular hermaphroditism" in which the ovarian
tissue had been removed at an earlier stage of development. They
can be interpreted most readily as instances of abnormal sexual
differentiation in the genotypic male, the following assumptions
being made:

(1) The stimulus to differentiation of the remainder of the sex-
equipment is, in the mammal, localised in the gonads.

(2) The abnormalities pertain only to the earlier stages of sexual
development.

[1] Hughes (1927), however, has recently adduced evidence which shows that
the free-martin does occur in the pig.

(3) The influence of the gonad in the mammal at this stage is such as inhibits the further development of these structures of the accessory sexual apparatus appropriate to the alternative functional sex, and these structures, in the absence of such inhibition, continue their development unchecked.

(4) There exists a different threshold of response to the sex-differentiating stimulus on the part of different structures of the sex-equipment and at different times during the development of one and the same structure.

It is recognised that in this differentiation other agencies than the sex-hormone are involved, for example, the other endocrines, differences in the threshold of response on the part of the same structure in different cases, and that for effective differentiation there must be appropriate nutrition; but for the present these are disregarded.

For purely descriptive purposes it is assumed that in the process of the differentiation of the sex-organisation in a genotypic male, excluding that of the gonads, there are three overlapping phases: (1) the modelling of the external genitalia; (2) the atrophy of the Müllerian ducts; and (3) the further development of the Wolffian duct derivatives (Fig. 25). For the sake of simplicity it is assumed that for all these structures there is one and the same minimum stimulus which, provided by the male sex-hormone elaborated by the testis, will evoke the specific response toward appropriate development. It is also assumed that when once the undirected development of any structure has proceeded for some time then that structure is no longer capable of responding to the stimulus, if and when this is exhibited.

In A the minimum stimulus necessary for proper differentiation is exhibited before the time for differentiation has been reached, and as a consequence the differentiation is such that a completely male organisation is established.

In B, in consequence of a retardation in the elaboration of the sex-hormone or of the production thereof at a slower rate, the differentiation of the external genitalia is partially undirected, and the result is grossly imperfect external genitalia in an otherwise normal male. The erectile organ will be affected most, for the scrotum of the normal pig is sessile, and if in an abnormal male the testes descend through the inguinal canals they will become accommodated in a scrotum very much, if not quite, like that of the normal.

In C, in consequence of a greater retardation or still slower production, the end-results will be external genitalia even more imperfect and Müllerian duct derivatives further developed. In D the external genitalia will be as in C, the Müllerian derivatives will be better developed, whereas the development of the Wolffian derivatives will not be so complete.

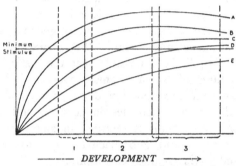

Fig. 25. Suggested application of Goldschmidt's hypothesis
to cases of intersexuality in mammals.

In E the required stimulus is never exhibited during the period of the differentiation of the accessory sexual apparatus and external genitalia and the end-result will be the full grown embryonic form.

In the absence of the proper endocrine control during the period of differentiation of the accessory sexual apparatus and external genitalia, these structures pursue a parallel development under the common stimulus of nutrition. Epididymes, vasa deferentia, and seminal vesicles develop from the Wolffian duct, uterus and vagina from the Müllerian, and if such undirected growth continues, then after a time these structures will have lost their embryonic plasticity, and will fail to respond to the stimulus of the sex-hormone, if and when this is exhibited later. The degree of development of the derivatives of the Müllerian ducts in a male and of the Wolffian ducts in a female, and especially the degree of imperfection of the external genitalia in a male, will provide significant indication as to the time during development at which the sex-hormone was exhibited.

These grades of abnormality of the reproductive system in the pig are thus to be explained as the result of the more or less complete absence, or of the qualitative or quantitative insufficiency, of the tissue in which the sex-hormone has its origin, during that period of development when the differentiation of the rest of the sex-equipment occurs. But since this differentiation of the gonad occurs before the time of the assumption of many of the secondary gonadic characters, these will be exhibited when that time is reached. So it can happen that an individual whose external genitalia, really nothing more than a full-grown urogenital sinus and genital tubercle, are indistinguishable from those of an immature female, may assume the secondary gonadic characters of the functional male and exhibit the exaggerated male sex instincts of a "rig." It is possible that the testis of the male and the structures of the mechanism concerned in its descent pursue a parallel and corresponding development and differentiation up to the point when descent occurs, and that if the differentiation of the testis is retarded in any way, the proper association of testis and gubernacular apparatus is prevented or embarrassed so that maldescent occurs.

The frequency of this type of abnormality in the goat and the pig, and the facts that it is more common in certain districts than in others, and that certain individuals in successive matings yield one or more of these intersexual offspring, point to the conclusion that this retardation or insufficiency of the sex-hormone is a character in the genetic sense.

In order to explain this retardation in the exhibition of the sex-hormone the following assumption is made. Different male-determining and female-determining gene-complexes elaborate their sex-differentiating substances at different rates or come into action at different times. The intersexual individual of this class is the result of the mating of individuals which transmit to their offspring that sex-determining gene-complex in which the factors are "slowly-elaborating" in nature, so that the minimum stimulus for differentiation of the gonad in the male is not exhibited at the critical time during development.

It is stated that after complete ovariotomy the Müllerian duct derivatives undergo considerable atrophy. In the intersexual pig the uterus is often very well developed indeed and yet testicular tissue alone is present. It will be found that if testicular grafts are placed in an ovariotomised female the uterus will not undergo atrophy. The action of the sex-hormone of the male is to inhibit the further development of the growing Müllerian ducts; it does not effect an atrophy of the fully developed uterus, in fact it would seem that the adult structures of the accessory sexual apparatus do not atrophy when either ovary or testis is present.

(b) *Intersexuality due to Abnormality in the Mode of Differentiation of the Gonad.*

The cases in class 2 are instances of as perfect intersexuality as is possible in the mammal. The essential feature of this condition is that both ovarian and testicular tissue shall be present synchronously or consecutively in one and the same individual. The mammal can and occasionally does possess both kinds of gonadic tissue, but it cannot function both as male and female since the external reproductive organs cannot be both male and female in their architecture. Moreover, oogenic tissue cannot flourish in a scrotum, nor spermatogenic within the abdominal cavity. Crew (1921) and Moore and Quick (1923) have shown that the optimum temperatures for efficient ovarian and spermatic functioning are widely different, and that the temperature within the abdominal cavity is considerably higher than that within the scrotum. Functional hermaphroditism is impossible, therefore, in the mammal, though possible in other vertebrate forms.

The cases examined by Crew and those recorded in the literature are similar to those of class 1 in every respect save that the gonads include both ovarian and testicular tissue. To interpret these cases of "glandular hermaphroditism" it is necessary only to explain the presence of both kinds of gonadic tissue, since the abnormalities of the accessory sexual apparatus and of the external genitalia can be explained most simply and yet quite satisfactorily by assuming that they are exactly the same in nature and in origin as those found in the intersexual male of class 1. These intersexes are to be found among the herds that produce the intersexual individuals of class 1; in one and the same litter both kinds of abnormal offspring may be produced.

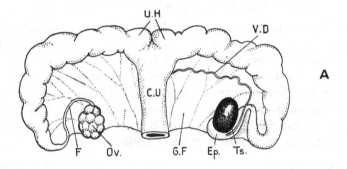

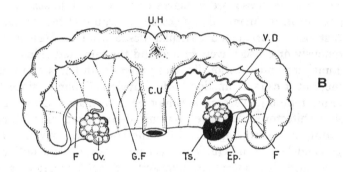

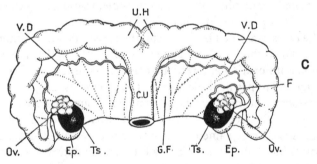

Fig. 26 A, B, C. Glandular hermaphroditism in the pig.

C.U.	Corpus uteri.	Ov.	Ovary.
Ep.	Epididymis.	Ts.	Testis.
F.	Fallopian tube.	U.H.	Uterine horns.
G.F.	Genital fold.	V.D.	Vas deferens.

In the pig the relation of ovarian and testicular tissues is remarkable, as has already been pointed out. The cases in which two ovaries are found within an abdominal cavity and two testes within an imperfect scrotum are to be interpreted as the result of the separation of ovotestes by the gubernacular apparatus. If an ovotestis is thus divided along the line of the connective tissue which invariably separates ovarian portion from testicular, two distinct gonads will be found on one and the same side of the body. If the testicular portion migrates to a situation beneath the skin of the perineum it may be expected that functional spermatozoa will be elaborated. But even were this the case, the imperfection of the external genitalia would not permit the individual to function as a male.

One point of considerable interest emerges from the study of these cases. If testicular tissue is present in one gonad, testis or ovotestis, an epididymis and vas deferens will be found associated with it, but if the other gonad is an ovary, no epididymis and no vas will be found on this side. There is more than hormone stimulation involved: it would seem that some mechanical stimulus for the maintenance of the epididymis and vas is supplied by a testis but not by an ovary.

The abnormality of the gonads can be explained if it is assumed: (1) that in these classes, males genetically, the sex-determining gene-complex included quickly-elaborating female-determining genes and slowly-elaborating male-determining genes; and (2) that the differentiation of the gonads is not synchronous but consecutive, the left being affected before the right, and the cephalad pole before the caudad. If these individuals are genetic males then as a consequence of the balance between male- and female-determining factors established at the time of fertilisation, sooner or later the male-differentiating reactions will be in excess, but if the male-determining genes are slowly elaborating and the female-determining genes are quickly elaborating then the situation will be such that the female-differentiating reactions will first exert the necessary minimum stimulus for gonadic differentiation and ovarian tissue will be laid down, and that shortly the male-differentiating reactions will overtake and replace the female, and the remainder of the differentiation will be pursued under their control, testicular tissues being laid down. The relative amounts of ovarian and testicular tissues will provide an estimate of the time during the period of gonadic differentiation when one kind of sex-differentiating reaction replaced the other. These suggestions can be graphically illustrated as follows.

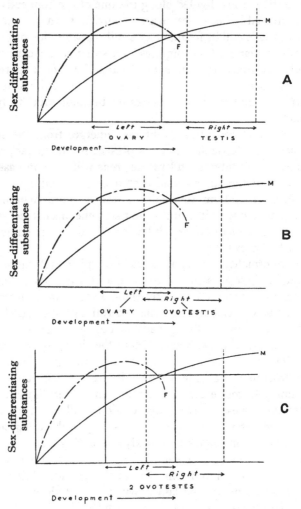

Fig. 27 A, B, C. Suggested interpretation of glandular hermaphroditism in the pig.

As a result of the simultaneous presence of ovarian and testicular tissues both male and female sex-hormones will be elaborated. It is assumed that this differentiation of the gonads has been retarded and that before it occurs the accessory sexual apparatus and external genitalia have assumed more or less the full-grown embryonic form. But this differentiation occurs before the time of the assumption of the secondary gonadic characters, and the structures concerned with these will be exposed to the action of both kinds of sex-hormone. The results are as those obtained in the experimentally produced hermaphrodite, a characterisation suggesting that certain structures respond preferentially to one sex-hormone, others to the other. The phallus invariably enlarges, while the sexual behaviour is exaggeratedly male.

If, on the other hand, the differentiation of the gonads, though abnormal in its mode, is normal in time, occurring before the differentiation of the rest of the sex-equipment is timed to take place, the situation can arise in which, owing to the presence of both ovarian and testicular tissue both types of sex-hormone are available for the direction of this differentiation. The external genitalia, in these circumstances, will be predominantly, if not entirely, male for the reason that the somatic tissues, being XO in constitution, will respond preferentially to the stimulus of the male sex-hormone. The internal genitalia will be a mixture of more or less well-developed derivatives of both Müllerian and Wolffian ducts.

Intersexuality can thus occur in two forms in the genetic male, one being the result of an abnormality in time of differentiation of the gonads, and the other of an abnormality in the mode of differentiation of the gonads. That it should be the male that is prone to such abnormality is not surprising, for the balance in the sex-determining gene-complex is known to be less stable in the digametic (XO) than in the monogametic (XX) sex.

Of course, it may well be that these individuals are genetic females (in fact, further investigation is providing evidence which would seem to show that they are); that the differentiation of the gonads is in the order right to left, caudad pole to cephalad, and that the female-determining factors are the relatively slowly elaborating. Baker (1925) is definitely of this opinion and has brought forward evidence which supports his view.

In the goat and also in the human being, judging from the descriptions of ovotestes given by different writers (*e.g.* Krediet, 1921), the course of gonadic differentiation is different from that which obtains in the pig, as is also the time relation in the differentiation of the structures of the accessory sexual apparatus and external genitalia. Nevertheless, with a few modifications, the scheme suggested for the pig can be adapted for the interpretation of the conditions found in the human intersexual, and of the very many hermaphrodites in other vertebrate forms.

Chapter IV

THE PHYSIOLOGY OF SEXUAL DIFFERENTIATION (*continued*)

B. INTERSEXUALITY DUE TO THE OVERRIDING OF THE GENOTYPE (PHYSIOLOGICAL INTERSEXUALITY)

(1) *Intersexuality through the Action of a Foreign Biochemical Agency.*

(a) *The Free-martin* (*zwicke*) *in Cattle.*
The part played by the sex-hormones in the production of this kind of intersexuality is best illustrated by the work of Keller and Tandler (1917) and of Lillie (1917) on the bovine free-martin, a genetic female (XX) co-twin to a normal male, the reproductive system of which becomes abnormal during the period of sexual differentiation as a result of the action of the sex-hormone of a male co-twin *in utero.* Twins may be monovular or identical, and these arising from one and the same egg are always of the same sex and very similar in their characters; or they may be binovular or fraternal, resulting from the synchronous fertilisation of two separate ova. In the latter case they may or may not be of the same sex and do not resemble one another any more closely than do brothers and sisters born at different times. Twins in cattle may consist of two normal males, two normal females, one male and one female each normal, or one male and the other an individual with an abnormal reproductive system and known as a "free-martin." In all save one of 126 cases of twins in cattle thoroughly examined by Lillie, two corpora lutea were found. This shows that twins are almost invariably binovular in this animal, since two ova are concerned in the pregnancy.

The two fertilised ova pass into the bicornuate uterus and become attached to the uterine mucosa. If they have been discharged from the same ovary, the zygotes usually develop in one and the same uterine horn. As the zygotes increase in size, the embryonic membranes of the two foetuses meet to adhere and in many cases to fuse. If such fusion occurs, an anastomosis of their blood vessels can result, so that a common vascular intercommunication may become established. It is to be noted that in the case of twinning involving one normal male and one normal female there is no such anastomosis.

Fig. 28. The bovine free-martin. (*After* Lillie.)

Thus the situation arises in which the sex-hormone of each developing individual is at liberty to pass into the tissues of its co-twin. The sex-hormone is the instrument which models the sex-organisation alone. The internal secretions of the pituitary, thyroid, adrenal, and so forth, can also pass from each individual to the other, but these are mainly concerned in the general and not in the special development of the individual and will be alike in both twins.

But if it so happens that the twins are bisexual (and a study of the sex-ratio in cattle shows that the laws of probability lead one to the conclusion that three possible sex combinations in twinning must occur in the proportion of 1 ♂♂ : 2 ♂♀ : 1 ♀♀, and that the twin combination that includes a free-martin must be regarded as male : female), and if the fusion of the chorions occurs, and further if a vascular intercommunication becomes established as it does in seven cases out of eight, the sex-differentiation of both individuals will be directed by that sex-hormone which is exhibited earlier or which is more potent. The testis becomes differentiated at an earlier stage of development than the ovary, and so the sex-hormone of the male is liberated before that of the female. The female twin (*i.e.* genotypic female) will pursue her sex-differentiation under the direction of the male sex-hormone of her co-twin and will therefore come to possess more or less completely the organisation of the male. The assumption of the male characters in the case of the foetuses examined is imperfect; the external genitalia are of the female pattern, the internal organs of reproduction more or less completely male. The male sex-hormone is liberated before the embryonic gonads of the genotypic female have undergone differentiation into ovaries; such differentiation is prevented and so there is no question of a competitive action between male and female sex-hormones. The end-result will have a relation to the time of exhibition of and to the efficiency of the male sex-hormone. The variation in the size of the testes of the male co-twin and of the extent of vascular intercommunication seems to point to the conclusion that the amount of the hormone is not a significant factor in the production of a free-martin but that there is a minimum stimulus and the reaction is of the "all-or-none" type. It is seen that the tissues of the genotypic male respond completely to the stimulus of the testicular sex-hormone, whereas those of the genotypic female fail to do so. Now it is known that genotypic male and female tissues are to be distinguished by differences in their chromosome content and it is reasonable to assume that because of these differences they differ also in their physiological constitution. It is probable, therefore, that though both kinds may be capable of responding to one and the same sex-hormone stimulus, they will respond differentially. The male co-twin develops testes because he is a genotypic male and becomes a phenotypic male because he develops testes. His embryonic gonads became testes because their differentiation was pursued under the direction of the male-differentiating reactions elaborated by the interaction of the genes in the sex-determining gene-complex. The sex-hormone elaborated by the testes, passing into the body of the genotypic female, swings her sexual differentiation in the male direction, but the swing is not complete because the sex-hormone of the testis is not equivalent physiologically to the male-differentiating substances elaborated by the sex-determining factors and because her tissues are constitutionally different from those of the genotypic male. The degree of development of the mammary glands of the free-martin is as that of the immature female. This fact is not without interest, for it is known that mammary development is not an infallible indication of the presence of ovarian tissue: in the human subject many cases of considerable development of the mammae in the male have been recorded. In the assumption of any character three variables are involved, the efficiency of the stimulus, the degree of response, and the sufficiency of nutrition. As has been stated, intersexuality is comparatively common in the goat: it is probable that many of these cases are free-martins, for Keller (1922) has shown that the same kind of placental anastomosis can be demonstrated.

That this is the correct interpretation of the case of the free-martin seemed to be supported by the work of Minoura (1921) who claimed to have produced an equivalent condition in the chick. However, Greenwood (1925), Kemp (1925) and also Willier (1925), repeating this work even more critically and more extensively, have completely failed to confirm Minoura's conclusions.

It is seen that in the mammal the most distinctive sex dimorphic characters are the secondary gonadic, and that for the development and maintenance of these the presence and action of functional gonadic tissue is necessary. The physiological action of gonadic tissues of the opposite sex causes the further development of incompletely differentiated structures to follow the direction appropriate to that sex and so far as is morphogenetically possible renders the individual intersexual. These conclusions are fully supported by the results of experimental intersexuality. (Steinach, 1916; Sand, 1923; and Moore, 1921.)

All the results of experimental embryology indicate that in the differentiation of the sex-equipment two phenomena are to be distinguished: (1) the development of the embryonic architecture, (2) the differentiation of the component structures during further growth, and the attainment of specific form under the direction of the genes resident in the chromosomes. The timeous production of these hormones is the function of the genes. In the insect and physiologically similar forms the products of metabolism that guide the development of tissues towards definite form and structure are present within the individual cell and are elaborated there almost, if not completely, independently of the rest of the body. In the mammal the control of differentiation is removed from the genotype of the individual cell to become the especial function of the glands of internal secretion. Intersexuality in the insect is the direct expression of an unusual genotype; in the mammal the genotype can be overriden.

The fecund tortoiseshell tom-cat is indeed a rarity and many have been the attempted interpretations of this undoubted fact. Little (1920) and Doncaster (1920) have each suggested that the infecund tortoiseshell tom is a feline free-martin, but Bamber (1922) who examined a series of multiple births found no indication of chorionic fusion or intervascular communication.

(b) *The Reciprocal Free-martin in the Opossum.*

Hartman and League (1925) describe a sex-intergrade opossum which was bought in as a male (for so it appeared), but on closer inspection it was found to have skin folds, simulating a pouch, better developed than in the normal male. The penis was of normal size and structure and the scrotum was well formed but empty; the head, previously described as being of the female type by Hartman, was of the male type. The internal genitalia consisted of an accessory sexual apparatus of the infantile female type. All the parts of the Müllerian duct were present as also were the vagina, lateral vaginal canals, uteri, and Fallopian tubes. The glands were infantile in dimensions and in structure. The round ligaments were normal and therefore large in proportion to the rest. There were no vasa deferentia and the Wolffian duct derivatives were absent. The gonads were in the position of the ovaries and were very small; their histological structure was somewhat indefinite.

Though the authors could not prove their contention that this abnormal individual was a male rendered abnormal whilst *in utero* by the action of the sex-hormone of a normal female co-twin, they present certain evidence in favour of this interpretation.

Whether the interpretation put forward is the correct one or not cannot be stated until further and fuller examination of similar cases has been made, but if it is granted that there is a chorionic anastomosis, that the ovary exerts its physiological activity before the testis, and that in the opossum there are secondary gonadic characters, there is no inherent flaw in the hypothesis.

(c) *The Case of* Bonellia.

The marine worm, *Bonellia viridis*, displays a remarkable degree of sex-dimorphism. The female has a plumpish green body about the size and shape of a plum and lives under stones or in a hole in a rock, its long slender terminally bifurcated ciliated proboscis protruding for food-catching purposes. The male is a microscopic pigmy whose internal organs, save those concerned with reproduction, are entirely degenerate and who lives as a parasite within the body of the female. The fertilised eggs hatch out as free-swimming larvae. If a larva settles down upon the sea bottom, it becomes, with few exceptions, and after a short period of sexual indifference, a female, but if by chance or perhaps through attraction it settles upon the proboscis of a female it becomes a male. The sexual fate of a larva is determined by an accident of position. Baltzer (1914) took larvae at various periods after they had settled upon a female but before they had become completely male and forced them to lead an independent life, and as a result he obtained intersexual forms, the degree of inter-sexuality varying with the length of time the larva had been allowed to remain upon the proboscis of the female. He stained the proboscis of the female with methylene blue and noticed that after the larvae had settled down upon it for two or three days they also showed the blue stain. It can be concluded that the larvae absorb material from the proboscis and that this is responsible for the arrest of growth and the direction of the sexual differentiation. The arrested pigmy male passes from the proboscis of the female into her mouth and then after a slight change he emerges there-from and ensconces himself in the reproductive duct through which the eggs pass out into the water.

Baltzer has more recently shown that weak solutions of the green skin of the female proboscis are poisonous to many of the smaller forms of aquatic life and that the pigmy males of *Bonellia* are extremely sensitive to weak solutions, over 1 in 3000 being lethal; mere contact with the skin produces no ill effect, it is the absorption of substances that is fatal. The wall of the reproductive duct in which the adult males live has no poisonous action.

The efficacy of purely external stimuli to influence sex-differen-tiation is also seen in the case of the slipper limpet, *Crepidula*,

which, introduced to this country from America, became such a
plague to oyster fisheries. *Crepidula* lives gregariously in chains;
the free-swimming young settle on older individuals and grow
where they settle. Each individual after attaching itself passes
through a phase of sexual indifference, next through a male phase,
producing spermatozoa, then through a hermaphroditic phase,
producing both spermatozoa and ova, and finally ends its days

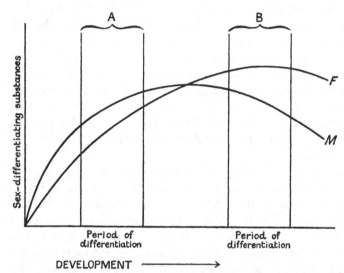

Fig. 29. Goldschmidt's interpretation of the case of *Bonellia*. A is the phase
of sex-differentiation pursued at a time during development when the
male-differentiating substances are effectively in excess; the result is a male
type of sex-organisation. B is the same phase pursued under conditions
when the female-differentiating substances are effectively in excess; the
result is a female type of sex-organisation.

as a female, producing ova only. Gould (1917) found that in
Crepidula plana the male phase occurs only if the individual
settles in propinquity to a larger individual, larger and older and
therefore a female; but that if the individual is isolated and
prevented from settling upon a female it passes directly from the
neutral to the female phase. A very similar state of affairs is
recorded in the case of *Crepidula fornicata* by Orton (1909).
An ingenious hypothesis has been advanced to cover the
observed facts concerning *Bonellia*, and in this matter the case

of *Crepidula* is very similar. It is suggested that the sex-determining genes produce sex-differentiating substances under the direction of which the differentiation of the sex-organisation is pursued. Goldschmidt (1923) supposes that in all the individuals there is at first an excess of male-differentiating substance, but that the production of the female-differentiating substance after a time overtakes this. Further, he supposes that the secretion of the proboscis of the female of *Bonellia* has the effect of accelerating the process of differentiation as opposed to the process of growth, of antedating, as it were, the period during development when sexual differentiation occurs. When differentiation is rapid, the sex-organisation matures under the influence of the male-differentiating substance; when it is not accelerated, under that of the female-differentiating substance. The mode of sex-differentiation is determined by a varying physiological state in connection with varying environment and secretions from other individuals.

Baltzer (1925) agrees in principle with this physiological interpretation of the case of *Bonellia* but holds that the cause of the intersexuality cannot be an acceleration of the rate of development but is rather a retardation. He points out that in experimental cultures of *Bonellia* there first appear normal females, then females with sperm, and lastly intersexes and males; and that moreover the male organisation, compared with that of the female, is to be regarded as a lower grade of development, being characterised by the absence of various organs.

(2) *Intersexuality through Parasitism.*

(a) *The Case of* Inachus *and* Sacculina.

Giard (1887) and later Smith (1906) have described in detail the changes that occur in crabs parasitised by *Sacculina* and other parasitic crustacea. *Sacculina*, an internal parasite, is a cirripede crustacean, and part of its body projects to the exterior under the abdomen of the crab in which it is living, while root-like processes, which absorb the juices of its host, ramify to all parts of the crab's body, avoiding the vital organs and absorbing nourishment chiefly from the blood. It attacks males and females and in both it causes atrophy of the sex-glands and consequent sterility. The only effect of this in the female is the acceleration in the assumption of the adult sex characters. Parasitised males, however, gradually take on more and more of the female characters, their great claws become smaller and smaller, the abdomen broader, the swimmerets enlarge and become fringed with the hairs to which, in the female, the eggs are attached. Most of the affected crabs die, but in a few the parasite disappears and the reproductive organs are regenerated. In a female a normal ovary develops, in a but partially feminised male, a normal testis, but in a fully feminised male a sex-gland is regenerated, in which both ova and sperm are found; real male intersexuality is produced. Geoffrey

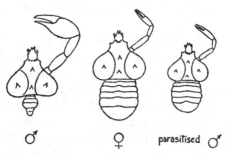

♂   ♀   parasitised ♂

Fig. 30. The case of *Inachus*. (*After* Smith.)

Smith (1906) who investigated this problem found that the blood of the normal female crab differs in chemical constitution from that of the male. It contains fatty substances which are absorbed by the ovaries and used in the production of the yolk of the egg. These fatty substances form an important part of the food of the parasite *Sacculina*. That which *Sacculina* absorbs cannot be used in yolk formation, and as the eggs cannot develop, the ovary degenerates. In the male these fatty substances are present in but small quantities. The parasite demands more and the whole physiology of the male crab is altered to meet this demand; the male thus assumes the female type of metabolism, and consequently the characters of the female.

Robson (1911) states that the infection by *Sacculina* induces the maintenance of an abnormal quantity of fat in the host liver and blood. This condition resembles that found in normal females and males preparing for the moult and in sexually mature females, the ultimate destination of the fat being functionally similar in the case of mature females and the infected crabs. In all probability the ultimate fate of the infected crab is death from starvation arising from its inability to obtain enough fatty material for itself and its parasite. A pink-coloured lipochrome is found in the blood of moulting and infected animals of both sexes, while a rich yellow characterises that of the sexually mature female.

The interest of this case is that it permitted Smith to question the validity of the conception that in all forms the gonads functioned as organs of internal secretion, contributing a peculiar product to the blood stream. Smith argued that the gonad, far from adding anything to the blood stream, removed something from it.

Goldschmidt, however, points out that it is not necessary to regard the change in metabolism as being the cause of the change in characterisation, for it itself can be regarded as a secondary gonadic character. It is well established that the blood of the male is chemically different from that of the female (Steche, 1912; Geyer, 1913; Manoilov, 1922–1923; Gräfenberg, 1922), whilst Farkas (1903), Strauss (1911), Lawrence and Riddle (1916), Lipschütz (1916) have shown that the metabolism in the two sexes is markedly dissimilar, but so far it has not been possible to induce intersexuality by means of blood transfusion.

Potts (1906) has shown that the infection of the hermit crab, *Eupagurus meticulosus*, by the cirripede *Peltogaster curvatus* has the effect of diminishing immediately the size of the gonads and of suppressing their functions, and that this is effected not directly but through interference with the general nutrition of the host. At an early stage of the external parasitism ova appeared in the glandular part of the testis: no corresponding changes were traced in the ovary. The male type of characterisation became transformed more or less towards the female type under the influence of parasitism. The same sequence of events was observed also in the parasitised *Eupagurus prideauxi*.

It is to be noted that the development of ovarian tissue in the parasitised male crab is actually a process of regeneration. It is of interest to note, therefore, that in the case of the worm *Ophryo-*

trocha puerilis (Braem, 1894) amputation of the genital segments of a female is followed by the development of testicular tissue in the regenerated portion. It is reasonable to assume that the conditions in which regeneration proceeds are responsible for the difference in the mode of differentiation.

(b) *The Case of* Thelia bimaculata.

Kornhauser (1916 and 1919) has shown that the male membracid *Thelia bimaculata*, when affected by a species of the genus *Aphelopus*, one of the Dryinidae, early in ontogeny assumes the female coloration. The pronotum of the male is normally dark brown with a bright orange-yellow vitta on each side; the female is grey, the vitta being only slightly visible. The males of *Thelia* possess in diploid number twenty-one chromosomes, the largest of which is the X-chromosome. The female has twenty-two, two of which are large X's. Females are larger than males and testes develop earlier than ovaries. Parasitised females are not affected. The degree of the change-over in the male depends on the state of development of the parasites in the fifth instar of *Thelia* previous to the final moult. The parasite, a hymenopteron, lays its eggs in the nymph of *Thelia*. Fifty to sixty larvae result and become full grown during the fifth instar of the host if the egg was deposited in a *Thelia* of the first or second instar. The larvae devour everything within the chitin of the host. If, however, they are only partially developed in the fifth instar they do not kill and the host becomes an adult but modified.

Early parasitism results in retardation of differentiation of external genitalia in both sexes but no transformation. Parasitism of the male is followed by considerable increase in the body size, while in the case of the female it is followed by a slight decrease. Parasitised males assume female-type wing size, size, pattern, and colour of head, size of proboscis, size of legs, size of digestive tube, and size of abdomen. Parasitism of females results in a decrease in body size but unaltered form save that the ventral and terminal plates are soft and non-pigmented.

The results of parasitism are not due to the initial effect on the gonad, for in a parasitised male the testes may be normal. The changes are to be associated with change in nymphal metabolism.

Male nymphs grow more rapidly, are smaller, darker and are sexually mature when they become adult, the testes filling the greater part of the abdomen. The male is katabolic. Female nymphs develop slowly and become larger, store fat, and are anabolic. Parasitism in the male brings about lower oxidation, storage of fat, retarded development, increased size. The external genitalia are laid down too early to be affected.

A somewhat similar case is that of the Andrenine bee parasitised by *Stylops* recorded by Pérez (1886).

(3) *Intersexuality through Gonadectomy.*

(a) *The Case of the Mammal.*

Pearl and Surface (1915) have described the case of a cow with cystic degeneration of the ovaries which assumed the general appearance of a bull. Rörig (1900) and others have recorded instances of female deer with ovarian disease developing horns or antlers as in the male. The difficulty that exists in such cases as these is that of distinguishing between male and female characterisations on the one hand, and the agonadic form on the other. Tandler and Gross (1913), for example, have shown that ovariotomy in the cow is followed by the assumption of a characterisation which is intermediate between those of the normal male and female.

(b) *The Case of the Fowl.*

The inter-relationship between gonad and plumage characterisation in the fowl can be demonstrated in the clearest possible way by an appeal to experimental gonadectomy and the implantation of gonadic tissues. The results of such experimentation are shown in the table on the opposite page.

The fact that the presence of both ovarian and testicular tissue in the male is associated with hen-feathering is of interest in connection with the case of the gynandromorph fowl described by Macklin.

It is established (Torrey and Horning, 1925; Cole and Reid, 1924; Crew, 1925, *et alia*) that administration of thyroid extract to cocky-feathered cocks is followed by the assumption of feathers very similar to those characteristic of the hen of the same breed.

Table III.

Sex		Effect on Plumage	Comb and Wattles
♂	Castration	Becomes as that of capon (male in colour, looser, more luxuriant). If a hen-feathered cock, it becomes as that of capon	Shrink
♂	Castration + implantation of testis[1]	Remains as that of a cock	Remain as those of cock
♂	Castration + implantation of ovarian tissue	After moulting becomes as that of the hen of the breed and variety to which he belongs	Become as those of hen
♂	Implantation of ovarian tissue	After moulting becomes as that of hen but more ruddy (warmer looking) (tail sickles intermediate)	Remain as those of cock
♂	Implantation of extra testicular tissue	Becomes as that of the hen of the breed and variety to which he belongs (tail sickles intermediate)	Remain as those of cock
♀	Ovariotomy[2]	Becomes as that of capon of the breed and variety to which she belongs	Shrink
♀	Ovariotomy + implantation of ovary	Remains as that of the hen of the breed and variety to which she belongs, so long as implant remains as ovarian tissue	Remain as those of hen
♀	Ovariotomy + implantation of testis	Becomes as that of a cock of the breed and variety to which she belongs	Become as those of cock
♀	Implantation of testis	Remains henny but more ruddy (tail sickles intermediate)	Become as those of cock
♀	Implantation of extra ovary	Remains henny (tail sickles intermediate)	Remain as those of hen

[1] The henny-feathered cock which is castrated and into which testis from a cocky-feathered cock is implanted, remains henny-feathered. The cocky-feathered cock which is castrated and into which testis from a henny-feathered cock is implanted, remains cocky-feathered.

[2] Incomplete ovariotomy is followed in certain cases by the assumption of male type plumage.

It may be expected that conditions of hyperthyroidism in the cocky-feathered cock will lead to the assumption of henny-feathering.

Cocky-feathering in those cases in which there is sex-dimorphism in the colour and structure of the plumage is commonly regarded as trustworthy indication that within the body there is, or was at the time when the plumage was developed, active functional testicular tissues: henny-feathering as an indication that there is or was active functional ovarian tissues. Gonadectomy in both sexes is followed, after a moult, by the assumption of a plumage which in its coloration is as that of the male of the variety to which the bird belongs (save that in particoloured breeds there is apt to be more white), whilst the barbules in the distal portions of the feathers of the hackle region are absent as in the normal male, but the plumage is much looser and far more luxuriant in its growth: the plumage characters of the capon and of the poularde are exactly alike. Since this is so, it is commonly argued that the gonads exhibit an endocrine function, the ovarian internal secretion possessing the faculty of inhibiting the development of the plumage characters of the male. The fact that in certain breeds the plumage of the cock is as that of the hen is explained on the assumption that in their functioning the testes of such a male are endocrinologically equivalent to the ovary of a hen. Such an interpretation is supported by the facts recorded in the above table, if it is assumed that in the case of a bird in which both ovarian and testicular tissues co-exist the internal secretion of the ovary is more potent or is produced in greater quantities than is that of the testes.

But there are several facts concerning the fowl which, while demonstrating a very definite inter-relationship between the gonads and the secondary gonadic characters, do not support the contention that the gonads function as endocrine organs. On the contrary, they point to the conclusion that the relation of gonads and secondary gonadic characters is similar to that which obtains in the case of *Inachus*. In this discussion it is most important to note that the plumage characterisation at the time of examination does not necessarily agree with the kind of gonadic tissue within the body at this time. Modification of the gonad does not affect feathers already grown. The plumage characterisation is an accurate reflection of the condition of the gonadic tissue at the time when this plumage was developed.

The cocky-feathered laying hen.

For example, there is no reason why a laying hen should not be cocky-feathered or why a functional cock should not have the plumage of a capon.

The cocky-feathered laying hen is to be explained by the fact that at the time when this plumage was developing the ovary had undergone a more than usually complete involution. The bird then for the time being was without a gonad and in the absence of the physiological activity of a gonad the plumage became as that of a capon; so the plumage of the poularde began its development. Then the ovary resumed its activity and the effect of its activity was to tighten up the plumage: it came into action too late to model its structure. The present writer has in his possession a laying hen that has been henny, cocky, cocky, henny, and is cocky once more. (Lippincott (1920) has described a very similar case.) The head furnishings, the attitude, the behaviour have always been as those of a normal hen.

The developmental capon and poularde. The capon-feathered functional male is the result of a delayed action on the part of the testis during development. Such males are extremely common among White Leghorn flocks. In a population of White Leghorn cockerels there will be found three types: one that becomes sexually active at a very early age, even at 5–6 weeks having large erect combs and tight plumage and attempting to crow and fight; another that develops as a bird that has been caponised within the first week of its life; its head furnishings remain bloodless, its body grows steadily in size to become long-legged and awkward, whilst its plumage is long and loose; and a larger class that falls between these two in respect of the attainment of sexual maturity. The breeder usually selects his males for breeding from amongst this class and these in their turn produce the three types of male. The capony class matures very late and in many cases that have come under the observation of the writer they have not attained full sexual maturity even in their third year, being developmental capons until they do. Examination will reveal testes of normal histological structure but in size equal only to those of the male chicken before the secondary gonadic characters have become expressed. Spermatogenesis is incomplete.

Occasionally a developmental poularde is encountered and is in every way similar to the developmental capon save that she is smaller and that her face is distinctly more like that of the hen. She possesses an ovary of normal structure but one that has remained immature. The fact that the female is smaller than the male, even in an agonadic specimen, would seem to suggest that perhaps size is largely determined by X-borne genes, these being duplex in the male, simplex in the hen.

A case that cannot be interpreted if it is postulated that the ovary and the testis of the fowl elaborate internal secretions specifically different is that recently described by Greenwood and Crew (1926). A Brown Leghorn female chick was ovariotomised when four days old and into her body were implanted the chopped-up testes of her brother. She became a typical masculinised female with large head furnishings and perfect cock plumage. At the age of seventeen months the bird moulted and the new plumage was as that of a normal Brown Leghorn hen. At first it was thought that this was due to absorption of the testis implant and regeneration of the ovary but as the head furnishings still remained as those of the male it was recognised that this could not be the case. Post-mortem examination revealed a very small amount of highly degenerate ovarian tissue and a mass of testis tissue much greater than that usually found in a cock of the same size.

The right gonad of the female, as is commonly the case in the ovariotomised hen (Domm, 1924; Bénoit, 1924; Pézard and Caridroit, 1923; Zavadovsky, 1922), had become differentiated as a testis in spite of the fact that the genotype of its constituent cells was XY.

At the time of the assumption of the adult plumage this hen became completely cocky-feathered. If no gonadic tissue had been present during the critical period of the development of the plumage this would have been cocky in colour and structure but loose and luxuriant as that of the capon. If at the beginning of this critical period gonadic tissue had been absent or insufficient the plumage would have started to develop the characterisation of the plumage of the poularde, but if shortly after this gonadic tissue of either kind, ovarian or testicular, had become sufficiently active, then, though in its coloration and structure the plumage would have remained unaffected, it would have become tighter and closer to simulate that of the cock. It follows then that at the critical time of development of the first adult plumage ovarian tissue was either absent or insufficient and that following this either ovarian or testicular tissue became physiologically sufficiently active, or else that ovarian tissue was either absent or insufficient throughout, but that sufficient testicular tissue was present either at the beginning or else immediately after this. If the bird was a poularde the plumage of which in the later stages of its development had been affected by the physiological action of ovarian tissue, then after a moult the plumage would have become completely henny, as indeed it did, the head furnishings would have always been as those of a hen of this breed and variety, which they were not, and only ovarian tissue would have been found post-mortem, which was not the case. It is reasonable to assume that the second adult plumage was developed under the stimulus of a mass of testis greater than that usually found in a bird of this size. If this is the case, it is necessary to explain the association of abundant testicular tissue and henny-feathering. It is to be noted that the testis tissue implanted was not that from a henny-feathered male such as a Sebright: it was from a Brown Leghorn, a breed in which the males are definitely cocky-feathered. The facts of the case can be accommodated if it is assumed that ovarian and testicular tissues, in respect of their own individuation, exert demands upon the general economy of the same kind but different in degree, that the functioning of an ovary is physiologically more expensive than is that of the testis, and that it is possible to supplement the demands of the testes so that they become equivalent to those of the ovary. This interpretation will explain the association of henny-feathering and both ovarian and testicular tissue as in Macklin's gynandromorph.

The assumption of male secondary gonadic characters by the senile hen. It is by no means uncommon for the senile female bird, wild or domesticated, to assume the plumage characters of the male of the breed and variety to which she belongs. A hen canary in her third year will cease to lay, play the part of the male in coitus, and begin to sing. Examination will reveal the fact that associated with this transformation there has been more or less complete destruction of ovarian tissue or a progressively diminishing physiological activity of this. Since, as has been shown, very similar results follow experimental gonadectomy, it is reasonable to interpret the phenomena as the result of pathological ovariotomy. Ordinarily during the succeeding years of the individual's life the oocytes in their growth make certain demands upon the general economy of the individual and maintain a metabolic level of femaleness, but, should the conditions be unfavourable for their growth as a result of the physiological exhaustion consequent upon excessive egg laying or from haemorrhage or tumour growth, then, in the absence of the inhibitory influence of the growing oocytes, the plumage becomes as that of a capon which closely simulates that of a male of the breed and variety to which the individual belongs. In old hens it is quite common for the head furnishings to increase in size and to become as those of the cock. In such cases it is found that the plumage remains henny. Examination will reveal the presence of degenerate ovarian tissue in which tumour growth has occurred. It would seem that there is a definite relationship between mitotic activity in the gonad and the size of the head furnishings.

The nature of the right gonad of the hen. Because it has been found that in the cases examined (Domm, 1924; Bénoit, 1924; Zavadovsky, 1922) the right gonad of the ovariotomised hen develops into a testis, it has been argued that the hen is a constitutional hermaphrodite, that the right gonad, which in the normal female chick ceases its development at or about the sixth to eighth day of incubation, is from its beginning a testis but that being a testis in the presence of an actively developing ovary, it is not permitted to flourish. It is difficult to maintain such a thesis as this and in the light of more recent work it is far more satisfactory to explain the fact that this incompletely atrophied gonad becomes a testis in most cases following ovariotomy on the assumption that in spite of the fact that its tissues possess the XY type of constitution they are to a very considerable extent ambivalent in respect of their future differentiation and that their mode of differentiation is decided by the kind of physiological environment in which they find themselves. It will be remembered that in the case of the male the products of the primary proliferation of the germinal epithelium proceed to the development of testis without check, whereas in the case of the female these products of the primary proliferation of the germinal epithelium cease their development at about the sixth day of incubation and are replaced by the products of a secondary proliferation which proceed to become differentiated, in the case of the left gonad, as ovarian tissue. It will be shown later that any subsequent proliferation of germinal epithelium leads, if it leads anywhere, to the production of testicular tissue. Ovarian tissue is developed solely from a proliferation of germinal epithelium which follows a degeneration of the products of a previous proliferation that occurs about the sixth day of incubation. It follows that the difference between differentiation into ovarian or into testicular tissues is a reflection of a difference not inherent in the tissues themselves but in the environment in which the original ambivalent tissue develops. If and when the internal environment which obtains at the sixth day of incubation can be analysed and reconstructed, it will be possible to direct the differentiation of the products of any proliferation of germinal tissue into ovarian tissue. In this connection it is of interest to note that Greenwood (1925) has found that, though in the majority of cases in which ovarian tissue has been implanted it ceases to develop as such but is replaced by testicular, in certain cases it can continue its development as ovarian. It is not without significance that in quite a number of female hawks and owls there is a right ovary with normal follicles.

From this it will be foreseen that if in a hen the ovarian tissue is destroyed and if there is a recrudescence of activity in the rudimentary right gonad or if a proliferation of germinal epithelium occurs, then testicular tissue will become differentiated, perfectly or imperfectly, and the phenotype as far as plumage, head furnishings, spurs, and behaviour are concerned will be transformed from that of the hen to or towards that of a cock. Greenwood (1925) found that no secondary proliferation of sex-cords occurred in ovarian grafts in cases of ovarian implantation made later than the fourth day after hatching. In the case of older fowls the conditions of the internal environment are such that it may be expected that no new ovarian tissue will ever develop but that testicular tissue may develop in any female.

(4) *Intersexuality due to the Impress of External Agencies.*

In the dioecious species of oysters three intersexual individuals have been recorded. Kellogg (1892) described one in the American oyster, *Ostrea virginica*, and Amemiya (1925) two in the Portuguese oyster, *O. angulata*. These latter two contained eggs and sperm. The eggs were of ordinary size and shape, being similar to those of other dioecious oysters and much smaller than the eggs of species that are customarily hermaphroditic. Since Amemiya found these specimens among animals which had been kept in the tank for a considerable period, he inclines to the conclusion that the intersexual condition was definitely related to changes in nutrition. In this connection the work of Orton (1921) on the native oyster, *O. edulis*, is of great interest as is also that of Gemmill (1896) and of Orton (1919) on the limpet.

Spärck (1925), working with *O. edulis*, found that in this form male and female states alternate. The frequency of the occurrence of the female state appears to be influenced by external temperature. At a temperature of 20–22° an individual assumes the female state once a year, at 14–16° only once in 3–4 years.

Nachtsheim (1923) records that in females of *Carausius* (Hymenoptera), the development of which proceeded at 25° C., the ventral portion of the thorax took on the red coloration characteristic of the male. This is possibly due to a change in the metabolic level.

Chapter V

SEX-REVERSAL IN THE ADULT INDIVIDUAL
(as opposed to transformation during differentiation)

By this is meant the transformation of the sexual characterisation of an individual from that which is normal in one functional sex to that which is normal in the opposite; it is the process by which an individual of a normally dioecious group that has functioned, or has been so equipped as to be capable of functioning, as a female (or a male), becomes transformed into one that can function as a male (or as a female). It is protandrous or protogynous hermaphroditism in an individual in which hermaphroditism is not customary. Intersexuality in many instances is incomplete reversal, as was seen in the case of *Lymantria*, and it follows therefore that sex-reversal in the adult will be produced by agencies of the same nature as those which evoke intersexuality during ontogeny, under conditions in which there is an appropriate stimulus towards sex-transformation and in which the organisation of the individual is capable of responding thoroughly to this stimulus.

There must be a switch-over from one type of metabolism to another, from male to female, or *vice versa*. This may be the result of a significant change in the internal environment itself, due to genetic causes (cf. *Lymantria*), or, in certain cases, to functional abnormality of one or more of the individual's own endocrine glands (cf. adrenal virilism in the fowl, Berner, 1923, and pituitary or adrenal disorder in man, Blair Bell, 1920). Such abnormality may be due to accidents of development or to disease (cf. parasitism in *Inachus*; pathological gonadectomy in the fowl).

On the other hand, the change in the internal environment may be the reflection of a significant alteration in the conditions of the external environment, changes in temperature or nutrition, for example. The agencies which provoke sex-transformation are thus of two sorts: those inherent in the genotype, and those which in their action override this.

In order that, under the changed conditions, sex-reversal may proceed to its conclusion it is necessary that the component structures of the sex-equipment must be capable of transformation or replacement, one kind of gonadic structure must be replaced by another, the ovary must be transformed into, or replaced by, a testis, or *vice versa*; the accessory sexual apparatus, the external organs of reproduction, the rest of the secondary gonadic characters including the epigamic must be capable of becoming remodelled or replaced.

It follows that sex-reversal cannot occur in any mature individual the external and internal genitalia of which are fashioned early in its life history and thereafter lose all embryonic plasticity and become unresponsive to any stimulus which, had it been exerted before they were differentiated, would have guided their differentiation. Nor can it occur in any case in which the differences between male and female architecture are based upon the differential development of two different sets of structures one of which, in either sex, undergoes complete atrophy. Nor can it occur in cases in which sexual dimorphism involves a differential mode of development of one and the same set of structures, *e.g.* the urogenital sinus and tubercle in the mammal. If these become scrotum and penis nothing can change them into vulva and clitoris.

Until controlled experimentation has been carried out it is impossible to define the cause of sex-reversal in many cases. All instances of this phenomenon will therefore be included in this section.

In Mammals.

It cannot be expected that complete sex-transformation will occur after sexual differentiation has once taken place for reasons given above.

In Birds.

Sex-reversal in this group is not impossible theoretically because the external genitalia in the two sexes are very similar and because experimentation has shown that the functioning gonad plays an all-important rôle in sexual differentiation, it being possible to masculinise a female and to feminise a male by means of appropriate gonad implantation. In order that sex reversal may occur two conditions and only two would seem to be necessary: (1) gonadectomy through disease, and (2) the development of new sex-cords. Since the latter must inevitably in the mature bird give rise, if they develop at all thoroughly, to testicular tissues, sex-reversal is only possible in the case of the hen. The hen may become a cock, but a cock cannot become a functional hen.

Crew (1923) described the case of a Buff Orpington 3½ years old, a good layer, and the mother of chickens. On examination it was seen that the head of this bird was somewhat male-like, for her comb and wattles were rather larger than those of the typical hen. At this time the bird exhibited the classical signs of early ovarian disease. She had ceased to lay in the preceding autumn, and had moulted; she had spurs 3 mm. long on the left leg, 2 mm. long on the right; the plumage was entirely henny. She crowed weakly, as one practising, and her sexual behaviour was indifferent.

Two months later the comb, wattles, and spurs had progressively increased in size, and the bird had begun to moult irregularly. The feathers of the neck, saddle-hackle, and tail, as they were renewed, were seen to be cocky in structure, and ultimately she had become entirely cocky-feathered, though she could never retain the tail sickles. The spurs were now about 1 cm. in length, the left one being slightly longer than the right, and the legs had assumed the red tinge which characterises the male of the Buff Orpington. In the following spring she was crowing lustily and with a challenging note, was readily attracted by hens which would squat on her approach, and the sexual act would be performed. The bird fought with any and every male in the yard and was gently courteous to the hens. In fact, only by one accustomed to poultry or when placing it alongside a "real" cock could it be told that this bird was different from a typical male. Its stance differed from that of a cock; the bird was shorter on its legs, which formed a different angle with the body.

The bird was placed with a hen, a virginal Buff Orpington, in a pen far removed from all other birds. This hen was laying; the eggs which she laid during the fortnight previous to her mating were incubated and found to be infertile. Every egg she laid after the mating was incubated. Her mate performed the sexual act daily; fluid passed into the cloaca of the hen was withdrawn and examined; a few living spermatozoa were identified in the fluid. The hen became broody and nine of her own eggs laid during the preceding eighteen days were placed under her. Two chickens were hatched; the other eggs were clear.

Post-mortem examination of the abnormal bird revealed, lying in the situation of the ovary, a rounded mass with its purple surface marked with raised areas of yellow. Incorporated in the dorsal aspect of this mass there was a structure exactly resembling a testis, whilst another similar in appearance was situated in the corresponding position on the other side of the body. On the left a thin straight oviduct could be identified having a diameter of 3 mm. in its widest part near the cloaca; paired vasa deferentia were clearly discernible.

On sectioning, the structure of the gonads confirmed the conclusion that they were functional testes in a phase of reduced activity. The tumour proved to be the ovary, almost completely destroyed by tubercular disease.

The bird had been up to the age of three and a half years an unremarkable hen; she had laid many eggs and raised many of her own offspring. Her history was known, since her owner kept but few fowls. She began to suffer from ovarian disease, which became recognisable. The disease was tuberculosis of the ovary which progressively removed the ovarian tissue and so produced the effects of pathological ovariotomy. But it would seem that this tumour growth in its effects so altered the general metabolism of the individual that the conditions favourable to the differentiation and growth of spermatic tissue were created. New sex-cords developed from the germinal epithelium, and spermatic tissue was differentiated both in the left gonad and also in the incompletely atrophied right. The bird became anatomically equipped to function as a male, for with the development of the testes the Wolffian ducts were apparently stimulated to form functional vasa deferentia and the cloacal apparatus of the male was developed. Synchronously with the replacement of the ovarian tissue by spermatic the oviduct atrophied. This bird functioned as a male and became the father of two chickens. If this is indeed a case of complete sex-reversal in the fowl, in which a genotypic female as a result of the disturbance of metabolism by tumour growth had become a phenotypic male, and if the fowl has the *Abraxas* type of sex constitution, the sex-ratio of the offspring of this bird and a normal hen should be 50 : 100, so

P_1 XY   XY

Gametes   X   Y   X   Y

F_1 XX   XY   XY   YY (an infertile egg or dead zygote)

         ♂   ♀   ♀

Of the two offspring of the mating, both typical Buff Orpingtons, one was a male the other a female. These were interbred and their progeny were typical Buff Orpington chicks of both sexes.

The histological study of the gonads in this case and in seven others showing different stages in transformation, demonstrated beyond doubt that the birds were originally hens, the ovaries of which atrophied at some period of life and were then invaded by peritoneal tissue. This tissue in some birds consummated development by giving rise to mature seminiferous tubules, and others produced undifferentiated epithelial cords which either continued to grow indefinitely, thus developing into tubules of an embryonic or immature type, or forming a malignant tumour. Traces of the proliferation of sex-cords from the germinal peritoneum were met with in most of the ovaries.

The cases displayed a consistent seriation illustrating the conversion of an actively functioning female into an actively functioning male, and into this series could be placed the cases described by other investigators. A fowl which previously had been equipped with the sex organisation of the female and had functioned as such can undergo such a transformation as to come to possess the sex organisation of the male, and actually to function as a male.

Ovarian tissue is replaced by testicular, and the type of differentiation of the rest of the sex equipment pursued under the direction
of the functional ovary gives place to that type which is pursued
under the direction of the functional testis. It is necessary to
bring these facts into line with the established principle of the
zygotic determination of sex in the fowl and with the more general
problem of the overriding of the sex-chromosome mechanism.

The sex-chromosome constitution of the fowl is such as to make
the male XX, the female XY (Stevens, quoted by Boring, 1923;
Hance, 1924; Shiwago, 1924): moreover, the indisputable evidence
of sex-linked inheritance affords strong reason for holding that
sex in the fowl is determined at the time of fusion of the gametes
by a sex-chromosome mechanism of the *Abraxas* type. In the case
of the pigeon Riddle (1917) has demonstrated the existence of two
sorts of egg, a high storage low metabolic type developing into
females and a low storage high metabolic type that yield males.
This being so, then, as in the case of *Drosophila*, it might be
expected that any agency disturbing the balance between the two
sets of sex-determining factors, *i.e.* those borne upon the X-chromosome and those borne on the autosomes, in the cytoplasm, or
possibly on the Y-chromosome, would produce a greater effect in
XY individuals, in this case the hen. The fact that one of these
fowls, kept under the closest observation, had during the first
three years of its life the appearance, behaviour, and functional
powers of a hen and then later assumed the attributes of a cock,
makes it clear that the type of sex-organisation and of the reproductive functioning of the individual are not irrevocably decided
by the sex-chromosome constitution. It is certain that an XY
individual—a genotypic female—can produce sperm just as
efficiently as it can produce ova.

In the case of the fowl it has been shown that there are successive
invasions of the organ by sex-cords derived from the peritoneum
(Fell, 1923). The histological appearances suggest that so long as
growing oocytes are present these invading sex-cords do not
develop further into functional germinal tissue, perhaps being
transformed into "luteal" cells. But in the absence of growing
oocytes these cords are apparently converted regularly into seminiferous tubules. It would seem that the physiological conditions

which in the female embryo at the time of differentiation of the sex-organisation induce the primitive germ cells to assume the characters of oocytes—and it will be remembered that these are laid down before birth in the fowl—no longer obtain in the mature bird, so that if what may legitimately be regarded as some inhibiting influence of the functional ovary upon the invading sex-cords be removed, as is the case in ovarian atrophy and disease,

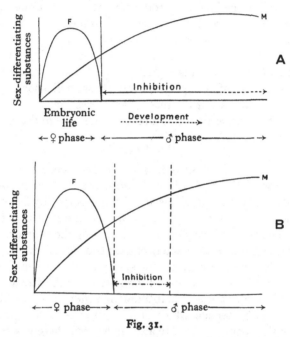

Fig. 31.

or, to put the matter differently, if by certain ovarian disease the conditions favourable for the continued development of the sex-cords are created, the germ-cells inevitably take on the characteristics of spermatogonia, spermatocytes and spermatozoa.

The phenomena of sex-reversal in the case of the female of the domestic fowl interpreted in terms of Goldschmidt's hypothesis can be illustrated as in Fig. 31.

During embryonic life the female-determining substances are effectively in excess and the differentiation of the gonad and the rest of the sex-equipment proceeds under the influence of the female-differentiating reactions: the oocytes are laid down. Ordinarily during the succeeding years of the individual's life the growth of the oocytes precludes the operation of the male-differentiating reactions which are increasing in efficiency (Fig. 31 A). But should the conditions be unfavourable for their growth, or should the conditions favourable for the continued development of the sex-cords arise as a result of the physiological exhaustion consequent upon excessive egg laying or from haemorrhage or tumour growth, then, in the absence of the inhibitory influence of the growing oocytes, the male-differentiating reactions become effective, spermatic tissue is differentiated, and the characters of the individual become as those of the male (Fig. 31 B). It can be expected that almost any hen of a highly fecund strain will sooner or later develop some degree of the male characterisation.

Riddle (1924) has recorded a very similar case in the ring-dove.

In Fish.

Essenberg (1923) records that sex-reversal occurs in the viviparous teleost fish, *Xiphophorus helleri* (the sword-tailed minnow), and that many instances of this have been reported by fish breeders and fanciers. In Essenberg's cases two females ceased to produce young when about three years old, and during the course of several weeks took on the sex-characters of the male. Cytological examination revealed the presence of ripe sperm in all parts of the gonad which, however, was juvenile in comparison to the size and age of the fish. Essenberg was able to show that there is a type of differentiation in the female which readily provides a morphological basis for the change-over. He also shows that there is a complete reversal of the sex-ratio in a population, this being 50 : 100 among immature fish and 200 : 100 among mature, a fact which supports the suggestion of sex-transformation of 50 per cent. of the females. In many males, moreover, the shape of the testes closely resembles that of the ovaries, and the grades suggest a transformation of pre-existing ovary into testis (but see van Oordt, 1925). It would seem from Essenberg's observations that sex-reversal is extremely common in this fish and that it is genetic in origin, there being a form which through genetic action is destined merely to pass through a female phase and later proceed to a male type of sex-differentiation. The case is very similar to that of *Lymantria*, save that even in a well-grown individual a remodelling of the sex-organisation can take place, there being no permanent hard parts.

Other cases of what would seem to be sex-reversal in fishes have been recorded by Herzenstein (quoted by Essenberg) in the cyprinodonts *Cymnocipris potanini* and *Schizopygopsis güntheri*, in which females assumed the sexual characters of the male. Philippi (1904) reported a similar case in the viviparous teleost, *Glaridichthys caudimaculatus*, and three others in *Glaridichthys januarius*, whilst Newman (1908) described a significant case of hermaphroditism in *Fundulus majalis*.

A very considerable number of cases of hermaphroditism in the different species of the fish have been recorded. It is unfortunate that in the majority of these the description is inadequate; however, the indications are that the conditions described are either stages in sex-reversal or are cases of intersexuality similar to those described in connection with the mammal (p. 84). That certain of them are indications of a sex-reversal probably equivalent to that in the case of *Lymantria* is very probable. In addition to the case of *Xiphophorus* there is the suggestive one of *Girardinus poeciloides*, the "millions" fish, examined by Huxley (1920). Boulenger had reported that the sex-ratio of a stock bred in the London Zoological Gardens was three females to one male for a period of nine to ten months, that this had then given place to a sex-ratio of one female to two males and lastly to one of equality. Huxley suggests that this swing in the sex-ratio is due to the fact that some of the females of the population in which the sex-ratio was three females to one male were genotypic males that had undergone sex-reversal to become phenotypic (or somatic as Shull, 1914, would call them) females. These phenotypic females, still XY in sex-chromosome constitution, would elaborate two sorts of eggs, in respect of the elements of the sex-determining mechanism, instead of one, and these being fertilised by X- and Y-bearing sperm would yield in every three on the average one female (XX) to two males (XY), the YY zygote being non-viable. If this sex-reversal had affected a certain proportion of the males of one generation, then two out of every three females would be normal in the next so that the sex-ratio would swing back to five females to six males. If sex-reversal continued for several generations there would ultimately be a swing in the opposite direction to give a preponderance of males.

Mršić (1923) has shown that in the trout over-ripeness of the egg (delayed fertilisation) is followed by sexual abnormality. He found that in the rainbow trout the gonads are first undifferentiated and that they then enter a phase in which the mode of differentiation is in the direction of an ovary and that following upon this there occurs, in a percentage of cases, a degeneration of the ova-like cells in the gonad which comes to assume the organisation of a testis. He was able to show further that the percentage of individuals exhibiting this change-over was related to the degree of over-ripeness in the egg at the time of fertilisation. The over-ripeness does not affect the chromosome mechanism but only the cytoplasm and the metabolic disturbance, he suggests, follows upon the fact that over-ripeness of the egg is associated with a resorption of germ cells of the ovarian type. It is probable that the case of the hermaphroditic trout described by de Beer (1924) was an isolated example of this transformation. In this case there was in a trout of 60 mm. a normal testis on one side, an ovotestis on the other; of this mixed gonad the ovarian tissue was cephalad, the testicular caudad. Mršić had observed during his studies of late fertilisation that not only do gonads change from ovary to testis but that at a certain stage of their development a testis may become transformed into an ovary. In seven trout reared from twenty-one days late fertilised eggs intersexual stages were present: the gonads were ovarian at the cephalad end, testicular posteriorly, and the change from ovary to testis was proceeding in the caudo-cephalad direction.

Langerhans (1876), Goodrich (1912), and Orton (1914) have recorded cases of intersexuality in *Amphioxus*. Orton's specimen was 4·4 cm. long and was predominantly male, only one of the 43 gonad pouches containing ova, the rest being filled with sperm. The digestive gland was also abnormal. Orton suggests that these cases are but stages in a process of sex-transformation in the direction male-female. It is to be noted that these intersexual specimens were of medium size.

IN AMPHIBIA.

Champy (1921) records the experimental sex-transformation in *Triton alpestris* Laur. A male, on being fed intensively after the winter's starvation, assumed the external characters of a female and instead of the testis there was an ovary as that of a young female. This case would seem to show that if at the end of a breeding season the gonads are physiologically exhausted, if the differentiated tissues undergo complete involution so that a new proliferation of germinal epithelium is required for the supply of gametes for the next season or if only a portion of the primordial germ cells develop each year, then the mode of the differentiation of each year's crop can be decided by the kind of environmental circumstances to which each in its turn is subjected. If also the male and female accessory sexual apparatus and external genitalia are very similar, or if, being developed from different rudiments which do not completely atrophy, rudimentary male ducts are present in the female and *vice versa*, and if the secondary gonadic characters, such as the crest of the newt, are developed afresh each year, then a complete sex-transformation would seem to be a relatively simple matter. All that is necessary is to control the conditions of the environment (temperature, food, etc.) during the critical period of the re-development of the gonads, and the rest follows.

Abnormality of the reproductive system in frogs and toads is relatively common. All grades of maleness and femaleness in the external sexual characters and in the gonads and accessory sexual apparatus have been recorded. Crew (1921) collected these cases and by arranging them in a series, at one end of which was a nearly complete female, at the other an almost perfect male, was able to put forward the suggestion that the gradations found were but stages in the process of complete sex-reversal probably from female to male. For reasons that could not be defined the ovaries of an adult female underwent degenerative changes and in the altered conditions spermatic tissue developed along the median border of the gonad. One of the specimens described had functioned as a male and of his progeny by a normal female every one was a female, as would be expected if this individual had been a genotypic female with the sex-constitution XX, which had under-

SEX-REVERSAL

gone a process of masculinisation. It has already been recorded that Witschi (1921) was able to show that such an hermaphroditic individual can elaborate both functional sperm and ova. In this case both sperm and ova were all X-chromosome-bearing and therefore yielded an abnormal sex-ratio in the next generation. Harms (1923) has found that if young castrated male toads are fed on a diet containing an excess of fat, lipoids, and lecithin for a considerable period of time the caudal portion of Bidder's organ becomes an ovary and the cephalic portion a new Bidder's organ. Oviducts and uteri are developed and the pointed shape of the head in the male is gradually transformed into the wide blunt head shape of the female.

BIBLIOGRAPHY (B)

ADLER, L. (1917). Metamorphosenstudien an Batrachierlarven. II. Der Einfluss überreifer Eier. *Arch. f. Entw.* 43, pp. 343–361.

AIDA, T. (1921). *Loc. cit.*

AMEMIYA, I. (1925). Hermaphroditism in the Portuguese Oyster. *Nat.* 116, p. 608.

BAKER, J. (1925). On the Descended Testes of Sex-intergrade Pigs. *Q.J.M.S.* 69, pp. 689–701.

BALTZER, F. (1914). Ueber die Bestimmung und den Dimorphismus des Geschlechts bei *Bonnelia*. *Sitzb. phys.-med. Ges. Würzb.* pp. 14–19.

—— (1914). Die Bestimmung des Geschlechts nebst einer Analyse des Geschlechtsdimorphismus bei *Bonellia*. *Mitt. zool. Stat. Neap.* 22, pp. 1–44.

—— (1925). Untersuchungen über die Entwicklung und Geschlechtsbestimmung der *Bonellia*. *Publ. Staz. Zool. Napol.* 6, pp. 224–286.

BAMBER, R. C. (1922). The Male Tortoiseshell Cat. *Jour. Genet.* 12, pp. 209–216.

BANTA, A. M. (1916). Sex-Intergrades in a Species of Crustacea. *Anat. Rec.* 11, pp. 505–506.

—— (1925). The Relation between Previous Sexual Reproduction and the Production of Male Offspring in *Miona macrocopa*. *Amer. Nat.* 59, pp. 50–61.

—— (1925). A Thelytokous Race of Cladocera in which Pseudo-Sexual Reproduction occurs. *Zeit. indukt. Abst.* 40, pp. 28–41.

BEER, DE, G. R. (1924). Note on a Hermaphrodite Trout. *Anat. Rec.* 27, pp. 61–62.

BELL, BLAIR (1920). *The Sex Complex.* London.

BÉNOIT, J. (1924). Sur la significance de la glande génitale rudimentaire droite chez la poule. *C. R. Acad. Sci.* 178, pp. 341–344.

BERNER, O. (1923). Virilisme surrénal chez une poule. *Rev. franç. d'endocr.* **1**, pp. 474–492.

BLUNCK, H. AND SPEYER, W. (1924). Kopftausch und Heilungsvermögen bei Insekten. *Zeit. wiss. Zool.* **123**, pp. 156–208.

BORING, A. M. (1923). Notes by N. M. Stevens on Chromosomes of the Domestic Chick. *Sci.* **58**, pp. 73–74.

—— AND PEARL, R. (1918). Sex Studies. XI. Hermaphrodite Birds. *Jour. Exp. Zool.* **25**, pp. 1–48.

BRAEM, F. (1894). Die Entwicklungsgeschichte von *Ophryotrocha puerilis.* *Zeit. wiss. Zool.* **57**, pp. 137–223.

BRANDT, A. A. (1889). Anatomisches und Allgemeines über die sogenannte Hahnenfedrigkeit bei Vögeln. *Zeit. wiss. Zool.* **48**, pp. 101–190.

BULLOCK, W. AND SEQUIRA, J. H. (1905). Relation of Adrenals to the Sexual Organs. *Trans. Path. Soc. Lond.* **56**, pp. 189–208.

CHAMPY, C. (1921). Changement expérimental du sexe chez le *Triton alpestris* Laur. *C. R. Acad. Sci.* **172**, pp. 129–134.

COLE, L. J. AND REID, D. H. (1924). The Effect of Thyroid Feeding on the Plumage in the Fowl. *Jour. Agric. Res.* **29**, pp. 285–287.

CREW, F. A. E. (1921). A Description of Certain Abnormalities of the Reproductive System found in Frogs and a Suggestion as to their Possible Significance. *Proc. Roy. Soc. Edinb.* **20**, pp. 236–252.

—— (1921). Sex Reversal in Frogs and Toads. *Jour. Genet.* **11**, pp. 141–181.

—— (1921). *The Cause of the Aspermatic Condition of the Imperfectly Descended Testis.* Thesis, Edinburgh.

—— (1922). A Suggestion as to the Cause of the Aspermatic Condition of the Imperfectly Descended Testis. *Jour. Anat.* **61**, pp. 98–106.

—— (1923). Studies in Intersexuality. I. (Mammal). *Proc. Roy. Soc.* B, **95**. pp. 90–109.

—— (1923). II. (Fowl). *Ibid.* pp. 256–278.

—— (1925). Rejuvenation of the Aged Fowl through Thyroid Medication. *Proc. Roy. Soc. Edinb.* **45**, pp. 252–260.

CUÉNOT, K. (1898). L'hermaphroditisme protandrique d'Asterina gibbosa et ses variations suivant les localités. *Zool. Anz.* **21**, pp. 273–279.

DOMM, L. V. (1924). Sex-Reversal following Ovariotomy in the Fowl *Proc. Soc. Exp. Biol. Med.* **22**, pp. 28–35.

DONCASTER. L. (1914). *Loc. cit.*

—— (1920). The Tortoiseshell Tomcat—A Suggestion. *Jour. Genet.* **9**, pp. 335–337.

ESSENBERG, J. M. (1923). Sex-Differentiation in the Viviparous Teleost *Xiphophorus helleri* Haeckel. *Biol. Bull.* **45**, pp. 49–96.

FARKAS, K. (1903). Beiträge zur Kenntniss der Ontogenese. III. *Pflüg. Arch.* **98**, pp. 490–546.

FELL, H. B. (1923). Histological Studies on the Gonads of the Fowl. I. The Histological Basis of Sex-Reversal. *Brit. Jour. Exp. Biol.* **1**, pp. 97–103.

132 BIBLIOGRAPHY

FINKLER, W. (1923). Kopftransplantationen an Insekten. *Arch. mikr. Anat.* 99, pp. 104–133.

GEMMILL, J. F. (1896). On Some Cases of Hermaphroditism in the Limpet (*Patella*), with Observations regarding the Influence of Nutrition on Sex in the Limpet. *Anat. Anz.* 12, pp. 392–393.

GEYER, K. (1913). Untersuchungen über die chemische Zusammensetzung der Insektenhaemolymphe. *Zeit. wiss. Zool.* 105, pp. 349–399.

GIARD, A. (1887). Sur la castration parasitaire chez l'*Eupagurus bernhardi* et chez la *Gabia stellata*. *C. R. Acad. Sci.* 104, pp. 1113–1116.

GLYNN, E. E. (1911–12). The Adrenal Cortex, its Rests, Tumours, its Relation to Other Ductless Glands and especially to Sex. *Quart. Jour. Med.* 5, pp. 158–192.

GOLDSCHMIDT, R. (1923). *Loc. cit.*

GOODRICH, E. S. (1912). A Case of Hermaphroditism in *Amphioxus*. *Anat. Anz.* 42, pp. 318–320.

GOULD, H. N. (1917). Studies on Sex in the Hermaphrodite Mollusc *Crepidula plana*. II. Influence of Environment on Sex. *Jour. Exp. Zool.* 23, pp. 1–70.

GRÄFENBERG, E. (1922). *Loc. cit.*

GRASSI, B. (1919). Nuove ricerche su la storia naturale dell' *Anguilla*. *R. Com. talassograf. Ital.* Mem. 47.

GREENWOOD, A. W. (1925). Gonad Grafts in Embryonic Chicks and their Relation to Sexual Differentiation. *Brit. Jour. Exp. Biol.* 2, pp. 165–187.

—— AND CREW, F. A. E. (1926). Studies on the Relation of Gonadic Structure to Plumage Characteristics in the Domestic Fowl. *Proc. Roy. Soc.* B, 99, pp. 232–240.

GUYER, M. F. (1909). On the Sex of Hybrid Birds. *Biol. Bull.* 16, pp. 193–198.

HANCE, W. (1924). The Somatic Chromosomes of the Chick and their Possible Sex Relation. *Sci.* 59, pp. 424–425.

HARMS, W. (1914). *Experimentelle Untersuchungen über die innere Sekretion der Keimdrüsen.* Jena.

—— (1923). Untersuchungen über das Biddersche Organ der männlichen und weiblichen Kröten. II. Die Physiologie des Bidderschen Organs und die experimentelle physiologische Umdifferenzierung von ♂ in ♀. *Zeit. Anat. Entw.* 69, pp. 598–629.

HARRISON, J. W. H. (1919). Studies on the Hybrid Bistoninae. IV. Concerning the Sex-Ratio and Related Problems. *Jour. Genet.* 9, pp. 1–38.

—— AND DONCASTER, L. (1914). On Hybrids between Moths of the Geometrid Sub-family Bistoninae. *Ibid.* 3, pp. 229–248.

HARTMAN, C. AND HAMILTON, W. F. (1922). A Case of True Hermaphroditism in the Fowl with Remarks on Secondary Sex Characters. *Jour. Exp. Zool.* 36, pp. 185–199.

—— AND LEAGUE, B. (1925). Description of a Sex-Intergrade Opossum with an Analysis of the Constituents of its Gonads. *Anat. Rec.* 29, pp. 283–292.

BIBLIOGRAPHY 133

HARTMANN, M. (1925). Untersuchungen über relative Sexualität. *Biol. Zentralb.* **48**, pp. 449–467.

HEGNER, R. W. (1914). *The Germ Cell Cycle in Animals.* New York.

HERTWIG, R. (1912). Ueber den derzeitigen Stand des Sexualproblems. *Biol. Zentralb.* **32**, pp. 66–129.

HEYMONS, R. (1890). Ueber die hermaphrodite Anlage der Sexualdrüsen beim Männchen von *Phyllodromia germanica*. *Zool. Anz.* **13**, pp. 431–437.

HUGHES, W. (1927). Sex-Intergrades in Foetal Pigs. *Biol. Bull.* **52**, pp. 121–136.

HUXLEY, J. S. (1920). Note on an Alternating Preponderance of Males and Females in Fish and its Possible Significance. *Jour. Genet.* **10**, pp. 265–276.

ISCHIKAWA, C. (1891). On the Formation of Eggs in the Testis of *Gebia major* de Haan. *Zool. Anz.* **14**, pp. 70–72.

KEILIN, D. AND NUTTALL, G. H. F. (1919). Hermaphroditism and other Abnormalities in *Pediculus humanus*. *Parasit.* **11**, pp. 279–288.

KELLER, K. (1922). Ueber das Phänomen der unfruchtbaren Zwillinge beim Rinde und seine Bedeutung für das Problem der Geschlechtsbestimmung. *Zentralb. Gynäk.* **46**, pp. 364–367.

—— AND TANDLER, J. (1917). Zur Erforschung der Unfruchtbarkeit bei den Zwillingskälbern des Rindes. *Südd. landw. Tierz.* **12**, p. 296.

KELLOGG, V. L. (1892). A Contribution to our Knowledge of the Morphology of Lammelibranch Molluscs. *Bull. U.S. Fish. Commiss.* pp. 389–436.

—— (1904). Influence of the Primary Reproductive Organs on the Secondary Sexual Characters. *Jour. Exp. Zool.* **1**, pp. 599–606.

KEMP, T. (1925). Recherches sur le rapport entre les caractères sexuels et les hormones des glandes génitales chez les embryos de poule. *C. R. Soc. Biol.* **92**, p. 1318.

KOPEĆ, S. (1912). Untersuchungen über Kastration and Transplantation bei Schmetterlingen. *Arch. f. Entw.* **35**, pp. 1–116.

—— (1922). Physiological Self-Differentiation of the Wing-Germ grafted on Caterpillars of the Opposite Sex. *Jour. Exp. Zool.* **36**, pp. 469–475.

KORNHAUSER, S. I. (1916). Further Studies on Changes in *Thelia bimaculata* brought about by Insect Parasites. *Anat. Rec.* **11**, pp. 538–540.

—— (1919). The Sexual Characteristics of the Membracid *Thelia bimaculata* F. *Jour. Morph.* **32**, pp. 531–636.

KREDIET, G. (1921). Ovariotestis bei der Ziege. *Biol. Zentralb.* **47**, pp. 447–455.

KUSHAKEVITCH, S. (1910). Die Entwicklungsgeschichte der Keimdrüsen von *Rana esculenta*. *Festschr. f. R. Hertwig*, **2**, pp. 63–224.

KUTTNER, O. (1909). Untersuchungen über Fortpflanzungsverhältnis se und Vererbung bei Cladoceren. *Intern. Rev. d. ges. Hydrobiol. u. Hydrographie*, **2**, pp. 633–667.

134   BIBLIOGRAPHY

LANGERHANS, P. (1876). Zur Anatomie des *Amphioxus lanceolatus*. *Arch. mikr. Anat.* 12, pp. 290–348.
LAWRENCE, J. V. AND RIDDLE, O. (1916). *Loc. cit.* (A).
LILLIE, F. R. (1917). The Free-Martin. *Jour. Exp. Zool.* 23, pp. 371–452.
LIPPINCOTT, W. A. (1920). A Hen which Changed Colour. *Jour. Hered.* 11, pp. 342–348.
LIPSCHÜTZ, A. (1916). Körpertemperatur als Geschlechtsmerkmal. *Anz. Akad. Wiss. Wien*, 53, Naturw. Kl. pp. 284–287.
—— (1924). *Loc. cit.* (A).
LITTLE, C. C. (1920). *Loc. cit.*
MANOILOV, E. O. (1922–23). *Loc. cit.*
MEISENHEIMER, J. (1909). *Experimentelle Studien zur Soma and Geschlechtsdifferenzierung.* Jena.
MINOURA, T. (1921). A Study of Testis and Ovary Grafts on the Hen's Egg and their Effect on the Embryos. *Jour. Exp. Zool.* 33, pp. 1–61.
MOORE, C. R. (1921). On the Physiological Properties of the Gonads as Controllers of Somatic and Psychical Characteristics. III. Artificial Hermaphroditism in Rats. *Jour. Exp. Zool.* 33, pp. 129–171.
—— AND QUICK, W. J. (1923). The Scrotum as a Temperature Regulator for the Testis. *Amer. Jour. Physiol.* 68, pp. 70–79.
MRŠIĆ, W. (1923). Die Spätbefruchtung und deren Einfluss auf Entwicklung und Geschlechtsbildung. *Arch. mikr. Anat.* 98, pp. 129–209.
NACHTSHEIM, H. (1923). *Loc. cit.*
NEBESKI, O. (1881). Beitrag zur Kenntniss der Amphipoden der Adria. *Arb. zool. Inst. Wien*, 3, pp. 110–162.
NEWMAN, H. H. (1908). A Significant Case of Hermaphroditism in Fish. *Biol. Bull.* 15, pp. 207–214.
OORDT, VAN, G. J. (1925). The Relation between the Development of the Secondary Sex Characters and the Structure of the Testis in the Teleost *Xiphophorus helleri*. *Brit. Jour. Exp. Biol.* 3, pp. 43–60.
ORTON, J. H. (1909). On the Occurrence of Protandric Hermaphroditism in the Mollusc *Crepidula fornicata*. *Proc. Roy. Soc.* B, 81, pp. 463–484.
—— (1914). On a Hermaphrodite Specimen of *Amphioxus*. *Jour. Mar. Biol. Ass.* 10, pp. 505–512.
—— (1919). Sex-phenomena in the Common Limpet. *Nat.* 104, pp. 373–374.
—— (1921). Sex-change in the Native Oyster. *Ibid.* 108, p. 580.
OUDEMANS, J. TH. (1899). Falter aus kastrierter Raupen, wie sie aussehen und wie sie sich benehmen. *Zool. Jahrb.* 12, pp. 71–88.
PEARL, R. AND CURTIS, M. (1909). Studies on the Physiology of Reproduction in the Domestic Fowl. III. A Case of Incomplete Hermaphroditism. *Biol. Bull.* 17, pp. 271–286.
—— AND SURFACE, F. (1915). Sex Studies. VII. On the Assumption of Male Secondary Characters by a Cow with Cystic Degeneration of the Ovaries. *Ann. Rept. Maine Agric. Exp. Sta.* pp. 65–80.

PÉREZ, J. (1886). Des effets du parasitisme des stylopes sur les apiaires du genre Andrena. *Act. soc. Linn. Bordeaux*, **40**, pp. 21–60.

PÉZARD, A. AND CARIDROIT, F. (1923). Les modalités du gynandro-morphisme chez les Gallinacés. *C. R. Acad. Sci.* **177**, pp. 76–77.

PFLÜGER, W. (1882). Ueber die geschlechtsbestimmende Ursachen und die Geschlechtsverhältnisse der Frösche. *Pflüg. Arch.* **29**, pp. 13–40.

PHILIPPI, E. (1904). Ein neuer Fall von Arrhenoidie. *Sitzb. Ges. naturf. Freunde, Berlin*, pp. 196–197.

POTTS, P. A. (1906). The Modification of the Sexual Characters of the Hermit Crab caused by the Parasite *Paltogaster*. *Q.J.M.S.* **50** pp. 599–621.

PRELL, H. (1915). Ueber die Beziehung zwischen primären und sekundären Sexualcharakteren bei Schmetterlingen. *Zool. Jahrb.* **35**, pp. 183–224, 593–597.

RIDDLE, O. (1916). Sex-Control and known Correlations in Pigeons. *Amer. Nat.* **50**, pp. 383–410.

—— (1917). The Theory of Sex as stated in Terms of Results of Studies on Pigeons. *Sci.* **46**, pp. 19–24.

—— (1924). A Case of Complete Sex Reversal in the Adult Pigeon. *Amer. Nat.* **58**, pp. 167–181.

ROBSON, G. C. (1911). The Effect of Sacculina upon the Fat Metabolism of its Host. *Q.J.M.S.* **57**, pp. 267–278.

RÖRIG, A. (1900). Über Geweihentwicklung. *Arch. f. Entw.* **10**, pp. 525–644.

SAND, K. (1921). Études expérimentales sur les glandes sexuelles chez les mammifères. I, II, III. *Jour. Physiol. path. gén.* **19**, pp. 305–322, 494–527.

—— (1923). L'hermaphroditisme expérimental. *Ibid.* **20**, pp. 472–487.

SCHMIDT-MARCEL, W. (1908). Ueber Pseudo-Hermaphroditismus bei *Rana esculenta*. *Arch. mikr. Anat.* **72**, pp. 316–339.

SCHÖNEMUND, E. (1912). Zur Biologie und Morphologie einiger Perlarten. *Zool. Jahrb.* **34**, pp. 1–56.

SCHREINER, K. E. (1904). Ueber das Generationsorgan von *Myxina glutinosa* L. *Biol. Zentralb.* **24**, pp. 91–104, 121–128, 162–173.

SEXTON, E. W. AND HUXLEY, J. S. (1921). Intersexes in *Gammarus chevreuxi* and Related Forms. *Jour. Mar. Biol. Ass.* **12**, pp. 506–556.

SHATTOCK, S. AND SELIGMANN, C. G. (1906). An Example of True Hermaphroditism in the Common Fowl. *Trans. Path. Soc. Lond.* **57**, pp. 69–109.

SHIWAGO, P. J. (1924). The Chromosome Complex in the Somatic Cells of Male and Female of the Domestic Chick. *Sci.* **60**, pp. 45–46.

SHULL, G. H. (1914). Sex-limited Inheritance in *Lychnia diocia* L. *Zeit. indukt. Abst.* **12**, pp. 265–302.

SMITH, G. (1906). Rhizocephala. *Faun. Flor. d. Golf. Napel.* Monog. No. 29.

SPÄRK, R. (1925). Studies on the Biology of the Oyster (*O. edulis*), in the Limfjord, with special reference to the Influence of Temperature on Sex Change. *Dan. Biol. Sta. Rpt.* **30**, pp. 1–84.

STANDFUSS, B. M. (1896). *Handbuch der paläarktischen Grossschmetterlinge.* Jena.

STECHE, O. (1912). Beobachtungen über Geschlechtsunterschiede der Haemolymphe der Insektenlarven. *Verh. deut. zool. Ges.* 22, pp. 272–281.

STEINACH, E. (1913). Feminierung von Männchen und Maskulierung von Weibchen. *Zentralb. f. Physiol.* 27, pp. 717–723.

—— (1916). Pubertätsdrüse und Zwitterbildung. *Arch. f. Entw.* 42, pp. 307–352.

STEINER, G. (1923). Intersexes in Nematodes. *Jour. Hered.* 14, pp. 147–158.

STRAUSS, J. (1911). Die chemische Zusammensetzung der Arbeitsbienen und Drohnen während ihrer verschiedenen Entwicklungsstadien. *Zeit. f. Biol.* 56, pp. 347–397.

SWINGLE, W. W. (1925). Sex-differentiation in the Bullfrog. *Amer. Nat.* 59, pp. 154–176.

TANDLER, J. AND GROSS, S. (1913). *Die biologischen Grundlagen der sekundären Geschlechtscharakteren.* Berlin.

TIHOMIROV, A. A. (1887). A Contribution to the Study of Hermaphroditism in Birds. *Trans. Imp. Soc. Nat. Hist.* Moscow, 52, pp. 1–29.

TORREY, H. B. AND HORNING, B. (1925). The Effect of Thyroid Feeding on the Moulting Processes and Feather Structure of the Domestic Fowl. *Biol. Bull.* 49, pp. 275–286.

—— (1925). Thyroid Feeding and Secondary Sex Characters in Rhode Island Chicks. *Biol. Bull.* 49, pp. 365–374.

VAULX, DE LA, R. (1921). L'intersexualité chez un crustacé cladocère (*Daphnia atkinsoni*). *Bull. Biol. Franc. Belg.* 55, pp. 1–86.

WILLIER, B. H. (1925). The Behaviour of the Embryonic Chick Gonad when transplanted to Embryonic Chick Host. *Proc. Soc. Exp. Biol. Med.* 22, pp. 26–30.

WINGE, Ö. (1922). *Loc. cit.*

WITSCHI, E. (1914). Die Keimdrüsen von *Rana temporaria. Arch. mikr. Anat.* 85, pp. 9–113.

—— (1915). Studien über die Geschlechtsbestimmung bei Fröschen. *Ibid.* 86, pp. 1–50.

—— (1921). Der Hermaphroditismus der Frösche und seine Bedeutung für das Geschlechtsproblem und die Lehre von der inneren Sekretion der Keimdrüsen. *Arch. f. Entw.* 49, pp. 316–338.

ZAVADOVSKY, M. (1922). *Sex and the Development of Sex Characters.* Moscow.

Chapter VI

THE MODE OF INHERITANCE OF
SEX-DIMORPHIC CHARACTERS

It has already been suggested that the sex-dimorphic characters are of two kinds: (*a*) the secondary genotypic, which are but the direct expression of the initial genotype, and (*b*) the secondary gonadic, which depend for their expression upon the physiological activity of the differentiated gonad. For purposes of discussion, it is convenient to speak of a third type of sex-dimorphic characters, viz. that based on a gene or gene-complex that is encouraged in its action by the conditions of an internal environment of maleness, and embarrassed by the conditions of an internal environment of femaleness, or *vice versa*.

Sex-dimorphism is evidenced when the sexes are distinguished by constitutionally different genotypes.

(1) In cases in which one sex has but one X-chromosome, the other two; when sex-linked characters are concerned; when the duplex condition of the X-borne gene yields a fuller expression of the character than does the simplex; and when the different characters "blend."

An instance of this type of sex-dimorphism is that provided by the multiple series of eye-colour characters in *Drosophila melanogaster*. Red (W), white (w), eosin (w^e), cherry (w^c), écru ($w^é$), tinged (w^t), blood (w^b), buff (w^{bu}), apricot (w^a), ivory (w^i), and coral (w^{co}) constitute this series. The results of mating any two of these are illustrated by the following example in which a white-eyed male is mated with a cherry-eyed female. Such a mating produces an F_1 in which the females have an eye-colour intermediate between cherry and white, and the males have cherry eyes:

P_1 Cherry ♀   (w^cX) (w^cX) × (wX) (Y)   White ♂
F_1 Cherry-white ♀♀ (w^cX) (wX) : (w^cX) (Y)   Cherry ♂

Similarly in the case of eosin and white:

P_1 Eosin ♀   (w^eX) (w^eX) × (wX) (Y)   White ♂
F_1 Eosin-white ♀♀ (w^eX) (wX) : (w^eX) (Y)   Eosin ♂♂

These eye-colours supply a reasonable explanation of certain forms of sexual dimorphism. In the case of animals it is not uncommon for the sexes to be strikingly distinct. The theory of sexual selection endeavours to explain the origin and perpetuation of the distinctive characters of the two sexes by suggesting that individuals of the one sex preferentially mated with individuals of the other which exhibited certain characters to a marked degree, and so *through time* the sexes diverged. But, as is seen in the case of cherry and eosin eye-colours, a striking difference may become established as a result of a single mutation. Many of the distinguishing sex-characters may be explained in this way— homozygosis and heterozygosis resulting from the XX and XY constitutions.

In those cases in which the X and Y chromosomes carry similar genes such sex-dimorphism in respect of sex-linked characters will not be exhibited. Such an instance has recently been recorded by Zulueta (1925) in the case of the beetle *Phytodecta variabilis*. Four colour phenotypes, striped, yellow, red and black, exist and it was found that striped is recessive to yellow, red and black; that yellow is recessive to red and to black; and that red is recessive to black. They constitute a series of multiple allelomorphs and of their genes only two of the same kind or one only of each of two kinds can be possessed by one and the same individual. These characters are sex-linked so that their genes are resident in the sex-chromosomes. The female is XX, the male XY. The results of breeding experiments demand that the Y-chromosome of the male shall carry one or other of the four genes. The fact that the same gene may be in the X and in the Y can be explained by assuming that in each chromosome an independent mutation had occurred, by postulating crossing-over between the X and the Y, as in the case of fish, or by calling upon the fairly common phenomenon in which a sex-chromosome is attached to an autosome so that of the apparent sex-chromosome only a portion is really the X or the Y, and by assuming further that crossing-over between the X and the Y portions does not occur, though it does between the autosomal components. Under these latter circumstances certain "autosomal" characters will exhibit partial linkage to sex and to truly sex-linked characters.

In certain instances it is not improbable that size differences between the two sexes are due to the fact that one sex is simplex the other is duplex for genes corresponding to sex-linked characters that affect the size of the individual.

(2) In cases in which one sex possesses the XY type of sex-chromosome constitution, the other the XX, and in which are involved characters the genes for which are Y-borne. The characters corresponding to the Y-borne genes will be sex-limited, being restricted to the digametic sex.

Examples of this type of sex-dimorphism have already been encountered in the cases of *Lebistes* described by Winge, and of *Aplocheilus* recorded by Aida. Castle (1922) has presented evidence which points to the conclusion that the abnormal character webbed-fingers in the human belongs to this category.

(3) Sex-dimorphism is evidenced when the sexes are distinguished by constitutionally different sexual genotypes so that in the one zygote an internal environment of "maleness" becomes established, one of "femaleness" in the other, and when in the genotype of both sexes there are genes that can exert their effects during development only in one or the other kind of internal environment, e.g. when there are genes that can only respond to the stimulus of "andrase" or of "gynase" but not of both. Such are the "sex-controlled" characters of Goldschmidt. They are sex-limited.

Instances of such sex-dimorphic characters are plentiful in forms in which it is known that the gonads play no part in the direction of sexual differentiation. Goldschmidt (1920) has pointed out that the males of the Hokkaido, Scheidemuhl and Aomori races of *Lymantria dispar* are to be clearly distinguished by differences in wing colour, and that crosses between these races show that these characters are transmitted in typical Mendelian fashion but are exhibited solely by the males. The females carry the genes for these characters, as breeding experiments show, but the characters are not expressed. However, in intersexual forms it can be seen that the females do exhibit the characters that distinguish their brothers. Goldschmidt has further shown that when the females of different races differ in respect of such sex-dimorphic characters, for example the yellow anal hairs of the Japanese races and brown-black hairs of the German races, crossing reveals typical segregation in F_2, the expected three types of female appearing (the third being the intermediate coloration of the F_1). An intersexual male of such a culture exhibits the character that distinguishes his sister.

Foot and Strobell (1914) working with the bug *Euschistus* record results that are readily interpreted in similar fashion. This interpretation is not that given by the authors but is preferable to it. The male of *Euschistus variolarius* has a black spot on its posterior end; the female lacks this. Neither sex in *E. servus* has this spot. When these two species are crossed, the males of F_1 have the spot, the females do not. In F_2 the females lack the spot, some males have it, some do not, whilst the rest present a graded series of intermediate conditions. There is no difficulty in explaining these results. The character spot is based on two or more multiple genes that give a summation effect; these genes are present in the genotype of *E. variolarius* but not in that of *E. servus*, and are only operative in an internal environment of maleness.

The phenomenon of unisexual polymorphism is also to be placed in this class of sex-dimorphic characters. Unisexual polymorphism, as the term implies, obtains when in a species all the individuals of one sex are alike in their characterisation whereas of the other sex there are several phenotypes. It can be illustrated by the case of *Papilio polytes* described by Fryer (1913) in which all the males are alike, whereas there are three different kinds of females, *cyrus* coloured like the male, *polytes* and *romulus* different from the male and from each other. Fryer's results were such as to show that these distinguishing characters were typically Mendelian and that for their interpretation the following justifiable assumptions were demanded:

A. A dominant gene modifying the *cyrus* coloration into *polytes* and only acting in an internal environment of femaleness.

B. A gene which in conjunction with A yields the *romulus* coloration:

	aabb = *cyrus* coloration	AABB	
	aaBb „ „	AaBB	
	aaBB „ „	AABb	
genotypes	Aabb = *polytes* coloration	AaBb	genotypes
of the females	AAbb „ „	AAbb	of the males
	AABB = *romulus* coloration	Aabb	
	AaBB „ „	aaBB	
	AABb „ „	aaBb	
	AaBb „ „	aabb	

Fryer found that all the results of his various matings could be accommodated by this scheme. Goldschmidt points out that the polymorphic sex in this and other cases examined at all fully is the XY sex, and that the results could be explained if there were a Y-borne gene for colour complementary to another auto-somal gene for colour. The critical test of this suggestion would be the examination of a case in which crossing-over between the X and the Y had occurred.

Male polymorphism, as encountered in *Parasemia plantaginis*, is to be interpreted in similar fashion and there is no theoretical reason why in one and the same species polymorphism should not be exhibited by both sexes.

Sex-dimorphism is evidenced when secondary gonadic charac-ters are present. Examples of this type of sex-dimorphism are exceedingly plentiful among the vertebrates though it has to be stated that in the majority of instances the relationship between the gonad and the characters concerned has not been experiment-ally proven. In some forms, however, the results of experimental gonadectomy and of gonad implantation have shown beyond all reasonable doubt that for the expression and maintenance of certain sex-dimorphic characters the physiological activity of the functional gonad, ovary or testis, is necessary.

Morgan (1915), and Punnett and Bailey (1921) have described the results of crossing a cock of a breed of fowls in which the plumage structure of the male is similar to that of the hen with hens of a breed in which there exists a marked sexual dimorphism in the matter of plumage structure, and also of castration of the hen-feathered cock. Hen-feathering in the male behaves as a typical Mendelian character and castration of the henny-feathered male is followed by the assumption of a capon's plumage.

In the presence of testes the plumage of the fowl (genotypic male or female alike) is cocky, unless the testis is that of a breed such as the Campine or Sebright in which both sexes are henny-feathered. In the presence of ovary the plumage is henny. In the absence of gonadic tissue of either sort the plumage is capony and similar in the castrated male and ovariotomised female. In a hemi-castrated henny-feathered cock the plumage is intermediate in character, and this kind of plumage, containing scattered cocky feathers among a henny ground, is not uncommon among the male progeny of the mating between henny and cocky-feathered breeds. If ovarian tissue is implanted into a cock, then in the presence of both ovarian and testicular tissue the plumage becomes henny; if testicular tissue is implanted into a hen, then in the presence of both kinds of gonadic tissue the plumage remains henny. It is difficult for the sex-hormone hypothesis to accommodate these varied facts and a simpler interpretation can be found. If it is assumed that the gonad, in the course of its own functioning, exerts a demand upon the general economy of the individual and that the demand on the part of the ovary is greater than that of the ordinary testis, as would be expected since the food storage of the egg is so much greater than that of the sperm, then ovary and testis in their functioning will establish distinct physiological states —there will be a male and a female type of metabolism—and this difference can be mirrored in the different plumage characterisation. It must be further assumed that the physiological demands of the testis of the henny-feathered cock are nearer akin to those of an ovary than to those of the testis of a cocky-feathered cock. Granting these assumptions, which can be tested readily enough by blood-fat and basal metabolism estimations, the facts outlined above can be reconciled. Gonadectomy removes all demands and the physiological level is the same in male and female. Hemi-castration of a henny-feathered cock lessens the demand, and if this coincides with the renewal of portions of the plumage there will be partial cocky-feathering. The supplementing of ovary with testis in a hen merely emphasises the demands of the ovary so that henny-feathering persists. The supplementing of testis with ovary in a male raises the demand, with the result that the plumage becomes henny. The intermediate character of the plumage of the hybrid from a cocky × henny mating is to be interpreted as the result of the bringing together of the two types of physiological con-stitution which in their inheritance obey the Mendelian scheme.

An interesting case concerning which all is not yet known is that of horns in the sheep. Darwin was already acquainted with the fact that when Suffolks (hornless in both sexes) were crossed with Dorsets (horned in both sexes), the F_1 hybrid generation consisted of hornless females and horned males. Wood (1909) has followed up the subject more exactly, and has shown that in the F_2 generation of this cross segregation takes place, giving ♀♀, 3 hornless, 1 horned; ♂♂, 3 horned, 1 hornless.

Bateson (1909) explained the result by assuming that the hornless race lacks a factor for horns which is present in the horned race, and this applies to both sexes. The presence of horns must then be regarded as recessive in the female and dominant in the male. This being so, the horn factor being represented by H and its absence by h, all the HH animals will be horned and all the hh animals hornless. Amongst animals simplex for H, however, the males will be horned and the females hornless. Morgan interprets the facts in another way: in the male the simplex state of H suffices to bring out the horn character whereas in the female the duplex condition of this gene is necessary. This explanation fits in quite naturally with the facts.

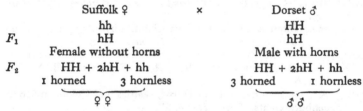

But it is clear that these formulae give no real explanation, and the same is equally true of others since put forward by Arkell and Davenport (1912), Goldschmidt (1913), and others. A real explanation necessitates the most exact knowledge of the relation of the internal secretions to horn growth. This is not yet available.

The evidence concerning the inheritance of horns in cattle is equally unsatisfactory. It is established that the polled condition is dominant and the horned recessive, but, as is seen in the experimental results of Watson (1921) and of Lloyd-Jones and Evvard (1916), the appearance of beasts with "scurs," loose button-like horns, in the F_1 and F_2 cannot as yet be completely explained on the assumption that the individual heterozygous for the polled character will develop scurs if it is a male. The mode of inheritance of horns in the goat has been shown by Asdell and Crew (1925) and also by Lush (1926) to be similar to that which obtains in the case of cattle.

BIBLIOGRAPHY (C)

ARKELL, T. R. AND DAVENPORT, C. B. (1912). Horns in Sheep as a Typical Sex-limited Character. *Sci.* 35, pp. 375–377.

ASDELL, S. A. AND CREW, F. A. E. (1925). The Inheritance of Horns in the Goat. *Jour. Genet.* 15, pp. 367–374.

BATESON, W. (1909). *Mendel's Principles of Heredity.* Cambridge.

CASTLE, W. E. (1922). The Y-chromosome Type of Sex-linked Inheritance in Man. *Sci.* 55, pp. 703–704.

FOOT, K. AND STROBELL, E. C. (1914). The Chromosomes of *Euschistus variolarius, Euschistus servus,* and the Hybrids of the F_1 and F_2 generations. *Arch. f. Zellf.* 12, pp. 485–512.

FRYER, J. C. F. (1913). An Investigation by Pedigree Breeding into the Polymorphism of *Papilio polytes* Linn. *Phil. Trans.* 204, pp. 227–254.

GOLDSCHMIDT, R. (1913). *Einführung in die Vererbungswissenschaft.* Leipzig.

—— (1920). Untersuchungen zur Entwicklungsphysiologie des Flügelmusters der Schmetterlinge. *Arch. f. Entw.* 47, pp. 1–24.

LLOYD-JONES, O. AND EVVARD, J. M. (1916). Inheritance of Colour and Horns in Blue-Grey Cattle. *Iowa Agric. Exp. Sta.* Bull. No. 30, pp. 67–106.

LUSH, J. A. (1926). Inheritance of Horns, Wattles and Colour in Grade Toggenburg Goats. *Journ. Hered.* 17, pp. 73–91.

MORGAN, T. H. (1915). Demonstration of the Appearance after Castration of Cock-Feathering in a Hen-feathered Cock. *Proc. Soc. Exp. Biol. Med.* 13, pp. 31–32.

PUNNETT, R. C. AND BAILEY, P. G. (1921). Genetic Studies in Poultry. III. Hen-feathered Cocks. *Jour. Genet.* 11, pp. 37–57.

WATSON, J. A. S. (1921). A Mendelian Experiment in Crossing Aberdeen Angus and West Highland Cattle. *Journ. Genet.* 11, pp. 59–67.

WOOD, T. B. (1909). The Inheritance of Horns and Face Colour in Sheep. *Jour. Agric. Sci.* 3, pp. 145–154.

ZULUETA, DE, A. (1925) La herencia ligada al sexo en el coleóptero *Phytodecta variabilis. Eos,* 1, pp. 203–231.

Chapter VII

THE SEX-RATIO

The sex-ratio is the numerical proportion of the sexes within a group. In biological literature it is commonly recorded as the number of males per hundred females: in biometrical papers it is expressed as the percentage of males in the data examined. The second method is greatly to be preferred for the reason that the probable error can be stated. A third and somewhat elaborate method is one that shows the proportion of males as a decimal of unity.

In any bisexual species there must be a sex-ratio at all times after the sex-determining mechanism has operated. For purposes of discussion it is convenient to take conception, birth, and maturity as the three salient points in the life-history of the individual at which to compute the sex-ratio of the species, and the proportions which obtain at these three stages are known as the primary, the secondary, and the tertiary sex-ratio respectively. Of these the secondary sex-ratio has received most attention.

Before the great mass of genetic and cytological evidence that is now available had been secured, the secondary sex-ratio and its variations were used as means of exploring the nature of the sex-determining mechanism itself. At the beginning of the present century it was generally accepted that at the time of fertilisation the egg was completely ambivalent as regards the future sex of the resulting zygote and it was customary to refer the sex of an organism to the conditions incident to development. It was known that the rudiments of the essential structures of the sexual equipment of both sexes were present in all embryos, and this fact was interpreted as an indication that all embryos were sexually completely ambivalent and that environmental agencies, particularly those of nutrition, could sway this initial neutrality towards either the male or the female type of sexual architecture as development and differentiation proceeded. However, it is now definitely established that in the majority of instances the sex-determining mechanism operates at the time of fertilisation, and that the genotypic sex of the individual is then finally and irrevocably determined: these more recent developments in genetics and cytology have made it clear that the use of the secondary sex-ratio as evidence regarding the nature of the agencies that determine sex was indeed a dangerous proceeding.

A. THE PRIMARY SEX-RATIO

It will be granted that in those forms in which one sex is digametic, the other monogametic, in respect of the elements of the sex-determining mechanism, the primary sex-ratio will be equality (1) if the two forms of gametes elaborated by the digametic sex are produced in equal numbers; (2) if these two forms are equally viable and functional; and (3) if fertilisation is at random. The primary sex-ratio is a demonstration of the sex-determining mechanism in operation and any modification of the ratio of equality indicates the influence of a directing agency upon that mechanism.

(1) It cannot be doubted that in many instances the primary sex-ratio is far from equality for the reason that the two forms of gametes elaborated by the digametic sex are produced in numbers that are far from equal. In the case of the bird and the moth the egg contains the X- and Y-chromosomes in conjugation before the polar bodies are formed. Into the first polar body goes either the X or the Y. If it is but a matter of chance which way this chromosome pair lies on the spindle, then equal numbers of X-bearing and Y-bearing eggs will result. But if in a particular line this pair should habitually be so orientated on the spindle that the X passes into the polar body more often than the Y, then in this line a preponderance of female offspring (XY) would be observed. Such a differential production can be obtained experimentally, as is shown by the work of Seiler (1920) on the Psychid *Talaeporia tubulosa* in which the female is the heterogametic sex. Seiler was able to show that the ratio of the eggs in which the X-chromosome passed into the polar body to those in which it remained in the egg, was exactly the same as the sex-ratio. Moreover, since in the course of these observations it was possible to detect the moment of the disjunction of the sex-chromosomes, it became possible to

attempt to influence this disjunction experimentally and so to disturb the sex-ratio. Seiler by varying the temperature during the maturation division obtained the following significant results:

Table IV.

Temperature	X-chromosome remained in the egg	X-chromosome passed into the polar-body	Sex-ratio
18° C.	61	45	136 : 100
35–37° C.	52	84	62 : 100
3·5° C.	48	31	155 : 100

Riddle (1916) submits that it is possible to distinguish between the two sorts of eggs in the pigeon, and has adduced a considerable amount of evidence in support of his contention. According to this authority, in pure species of pigeons the first egg laid, smaller and containing less chemical energy than the second, is destined, when fertilised, to become a male, the second a female. Cuénot (1899) and Cole and Kirkpatrick (1915), however, disagree with this conclusion.

Guyer (1909) had collected data on species hybrids among certain birds and had shown that there was a decided excess of males in the F_1 generation. The cause of this has not been demonstrated. Riddle (1916) recorded an excess of females in the cross *Streptopelia risoria* × *S. alba* (doves) under certain conditions and concluded that this excess was the result of a transformation of some of the males. It has been shown, however, that this conclusion is not justified for the cross involved a sex-linked character and of the hybrids the males are dark, the females white in colour, and examination of the data for sex and also for colour shows that the only possible explanation of the excess of females is that which postulates that the conditions of the experiment were such as to cause the X-chromosome to pass into a polar body at the time of the reduction divisions more often than to remain in the egg.

The quantitative difference that exists between the X- and the no-X or the Y-chromosome-bearing spermatozoa in many mammals, for example in the mouse, rat, pig, bull, horse, and man, may possibly supply an explanation of the differential production of the two sorts of gametes by the heterogametic sex; but it is likely that a disturbed primary sex-ratio is more commonly the result of a differential activity, susceptibility, or mortality on the part of the two sorts of sperm. It is necessary and not unreasonable to suggest that the Y-bearing sperms of the mammal possess some advantage during the period coitus-fertilisation. It may be that the male-determining sperm is the more active. If this is so, there exists the possibility of artificially separating the two kinds, the X- and the Y-bearing, or of further handicapping one or other by treatment of the female passages. In the case of the hetero-gametic female, environmental conditions, as is seen from Seiler's work, can cause a differential production of the two sorts of eggs as the result of a differential maturation division. Conditions may be such as to induce, for example in the case of the first egg of the pigeon, the female-producing genetic complex (= the Y) to be extruded into the polar body more often than the male-pro-ducing genetic complex (= the X), whilst as a result of further egg-laying, the conditions become such that the male-producing genetic complex is more frequently extruded. There is evidence which seems to show the existence of a differential chemical con-stitution in the early and the later egg, and it seems probable that this difference is mirrored in the different reaction systems in the dividing oocytes. Riddle has found that by forcing the female bird to lay eggs at more than the normal rate the proportion of females among her progeny is increased; the work of Pearl (1917) yielded suggestions that this might also be the case in the fowl, for signifi-cantly more females were produced by hens which had laid very heavily immediately prior to mating.

There is evidence which seems to show that polygyny in the mammal affects the sex-ratio. In the case of the human, whilst Sanderson (1879) and Campbell (1870) concluded that polygyny had no effect, Thomas (1913) gives figures that seem to indicate that the proportion of males is correlated with the number of wives in the case of the Ibo of Awka. But the degree of polygyny is so slight that these figures can at most be merely suggestive. In the case of the horse one stallion may serve a hundred or more mares during the course of a season and the number will vary with the reputed value of the male as a stock-getter. Düsing (1884–1887) gives the following figures for the horse which seem to indicate that the more the horse is used the higher is the sex-ratio.

Table V.

Number of times at stud	Offspring		Sex-ratio
	♂♂	♀♀	
60	71,407	70,569	101·19
55–59	75,493	74,912	100·77
50–54	69,972	71,461	97·92
45–49	69,774	72,073	96·81
40–44	66,573	69,045	96·42
35–39	44,911	46,493	96·60
20–34	29,023	29,934	96·94
	427,153	434,487	98·31

In the case of the mouse Parkes (1925) found that in monogamous matings the sex-ratio was 53·2 ± ·81, and that in seven lots of one male and 8–12 females the number of offspring was 395, the number of males 234, of females 161, and the sex-ratio 59·2 ± ·67, suggesting that polygyny raises the percentage of males.

It is probable that the effects of forced egg-laying in the case of the digametic female and of polygyny in cases in which the male is digametic are to be interpreted as being reflections of the fact that the production of one kind of gamete by the digametic sex is physiologically more expensive than that of the other. If the demand for the elaboration of gametes becomes a tax upon the gametogenic apparatus, then the kind of gamete that is elaborated more easily is elaborated in greater number (or the type that is more expensive is but roughly made and so remains handicapped when competing with the other kind). It is the X-bearing gamete of the digametic sex that is lacking either in number or in physiological perfection so that the overworked female bird provides more Y-bearing eggs, and the overworked mammal more Y-bearing sperm, with the result that the former yields more females, the latter sires more males.

It is well to bear in mind, however, the possibility that the differential production of gametes may be characteristic of a particular individual or strain, the result of some heritable mutation affecting the maturation division of the egg or leading to the suppression of one sort of sperm. This probably lies at the basis of such results as those of King (1918) in her inbreeding experiments with rats, in which she was able to shift the sex-ratio in either direction.

Bearing upon this question is the fact that in the aphids the no-X-bearing gametes degenerate; that in the male bee three out of the four gametes fail to develop; and that in *Hydatina* one of the primary spermatocytes yields two female-producing gametes, the other spermatocyte, failing to divide, degenerates. In all these cases fertilisation yields only females. If in the case of a form in which the female is the digametic sex there is a heritable mutation present that swings the maturation division always in one direction, as perhaps was the case in Doncaster's (1914) female line of *Abraxas*, or in the case of a form in which the male is the digametic sex, one that suppresses one kind of sperm as in the aphids, mentioned above, and perhaps in mammals also, a profound disturbance of the primary sex-ratio will be caused. The same will be true in the case of parthenogenetic forms in which one or other type of reproduction, with or without reduction, can be enforced.

The sex-ratio is affected by the relative degree of maturity of the ovum at the time of fertilisation in those cases in which the female is the homogametic sex. It is possible experimentally to modify profoundly the sex-ratio, as seen in the case of the frogs of Hertwig (1912) and of Kushakevitch (1910), or of the trout of Mršić (1923), and this may be explained on the assumption that the over-ripeness of the ovum is associated with extrusion of the X-chromosome into the polar-body, thus leading to the production of males. Hertwig allowed a male frog to fertilise half the eggs of a female and removed him from the nuptial embrace. The female does not lay her eggs in the absence of the male and so after an interval of any duration the male can be put back and will then fertilise the remaining eggs. The sex-ratio of the frogs hatched from the first half of the eggs was in every case near equality; with the rest the degree of disturbance of the sex-ratio varied with the length of the interval. Kushakevitch, Hertwig's pupil, repeated and confirmed these results.

Kushakevitch (1910) showed quite definitely that this result was not due to a differential mortality and Hertwig has described an actual histological remodelling of the gonad in these cases, showing

Table VI.

Hours...	0	6	18	24	36	42	54	64	89
				Per cent. males					
Hertwig ...	58	54	—	55	—	—	—	—	—
	49·0	—	—	—	58	—	59	—	—
	48·5	—	37	—	—	58	—	88	—
Kushakevitch	55	—	—	—	—	—	—	—	100

that in the late-fertilised zygote the gonad develops in a way quite distinct morphologically from that seen in the normal male. Bearing on this point is the observation of Adler (1917) that the thyroid of the late-fertilised zygote is markedly hypertrophied, which, if confirmed, may throw some light on the physiology of sex-differentiation.

These results can also be explained on the assumption that 50 per cent. of the eggs were fertilised by X-bearing sperm yielding XX individuals which normally would have become phenotypic females but that the conditions of the experiment were such as to transform them into phenotypic males, the sex-chromosome constitution in its action being overridden by the effect of the delayed fertilisation upon the egg's metabolism. If this were so, then functioning as males and fertilising the eggs of normal females they would produce offspring with a profoundly disturbed sex-ratio of two females to one male and this combined with the sex-ratio of offspring of normal parents, one female to one male, would give a preponderance of females, three females to one male, but all the individuals of this generation will be normal in genetic constitution and so will yield a succeeding generation exhibiting a normal one to one ratio.

Morgan (1919) suggests that Hertwig's and Kushakevitch's results may be due to the parthenogenetic development of over-ripe eggs.

If in a group there is a sex-chromosome sex-determining mechanism (e.g. XY : XX) and if this mechanism can be over-ridden, then in that group it can be expected that there will be individuals genotypically of one sex, phenotypically of the other, and these when mated to individuals in which sexual genotype and phenotype are in agreement will produce offspring among which there will be a preponderance of the sex to which the transformed parents genotypically belonged (*i.e.* a transformed female functioning as a male will yield a preponderance of females: a transformed male functioning as a female will yield a preponderance of males). If this process of transformation affects the individuals of several generations and thereafter ceases to act, there will be a decreasing preponderance of one sex followed by a preponderance of the opposite sex in the first generation after the close of the period during which sex-transformation has occurred and finally a sex-ratio of equality.

(2) A disturbed primary sex-ratio may be the result of a differential motility and vitality of the X and Y spermatozoa, or of a differential attraction for sperm by the X- and the Y-eggs of the heterogametic female. Such a conception as this postulates that there is a competition among the sperm. Cole and Davis (1914) have shown that this is the case. A rabbit was served by two dissimilar bucks, and it was found that the majority of the offspring claimed one of these bucks as their sire. In repeated matings this was always so. But when the sperm of this buck were alcoholised they could not compete with those of the other buck, though it was shown that when employed alone they could and did fertilise ova. It is reasonable to assume that if differences in the size of the sperm are associated with differences in motility, activity, or resistance to unfavourable conditions within the genital passages of the female, then chance would favour fertilisation by one rather than by the other kind of sperm.

The results of an experiment with fowls bear upon this question of competition among sperm. Crew (1926) placed two genetically dissimilar cocks with six hens. The great majority of the offspring claimed male A as their sire. The two males were then placed apart, each with six hens for ten days until these were laying fertile eggs, and then the males were exchanged. The influence of male A endured for nine days, that of male B for five days only. These results can be interpreted on the assumption that the sperm of male B became "stale," *i.e.* lost their vitality, more quickly than did those of male A.

B. THE SECONDARY SEX-RATIO

The secondary sex-ratio in a group will be identical with the primary sex-ratio if during the period conception–birth there is no sexually selective mortality among the zygotes. Any differences between primary and secondary ratios will serve as an indication of the prenatal survival value of the sexes.

A very considerable mass of data relating to the secondary sex-ratio, particularly in the case of mankind, exists, and a review of this will show at once that though fairly constant and not far from equality, the secondary sex-ratio is distinctly variable and rarely coincides with the standard of equality.

1. The secondary sex-ratio varies with the species. The secondary sex-ratio of most mammals has not been determined experimentally, for with few exceptions the mammals are unsuitable for experiments on a large scale, and that which is known has been culled from different Breed Registers. The following data have met with general approval (Cuénot, 1899; Schleip, 1913; Lenhossék, 1903; Hertwig, 1913; Doncaster, 1914; Morgan, 1911, 1914; Bonnier, 1923; Mohr, 1923; et alia):

Table VII.

Man	...	103–107	100	Mammalia
Horse	...	98·3	100	,,
Dog	...	118·5	100	,,
Cattle	...	107·3	100	,,
Sheep	...	97·7	100	,,
Pig	...	111·8	100	,,
Rabbit	...	104·6	100	,,
Mice	...	100–118	100	,,
Fowl	...	93·4–94·7	100	Aves
Pigeon	...	115·0	100	,,
Cottus	...	188·0	100	Teleostei
Lophius	...	385·0	100	,,
Loligo	...	16·6	100	Cephalopoda
Octopus	...	33·3	100	,,
Latrodectes	...	819·0	100	Arachnoidea
Lucilia	...	95·13	100	Diptera
Drosophila	...	100·0	100	,,
Macrodactylus		131·0	100	Coleoptera

2. The secondary sex-ratio varies with the race, breed, and strain. (a) Although the actual figures for a population considered

in the aggregate have but little biological significance, it is of interest to examine the figures of Bälz (1911) for the chief nations.

Table VIII.

Sex-Ratio of Entire Populations.

Country	Males per 100 females	Country	Males per 100 females
Great Britain	93·5	Belgium	98·4
Norway	94·0	Italy	99·0
Denmark	94·5	Poland	100·5
Sweden	95·3	Greenland	101·5
Spain	95·3	Japan	102·0
Austria	96·6	India	104·1
Germany	96·9	Bulgaria	104·5
European Russia	97·2	Serbia	106·0
Switzerland	97·2	Siberia	106·0
Hungary	97·7	Caucasus	111·0
France	97·9	Korea	113·0
Holland	98·2	Asiatic Russia	117·5
Ireland	98·3	China	125·0

(b) The Jewish race is remarkable for a high secondary sex-ratio which, in different areas and according to different authorities, ranges from about 140 to 105 males per hundred females, a ratio that indeed is high when compared with that for the general population of a particular area.

Table IX.

Sex-ratio of Jewish Race.

Place	Sex-ratio	Authority	Date
Livonia	139·84	Carlberg and Lenhossék	1873
Austria	138	Jacobs (1891)	1861–1870
Livonia	120	Darwin (1871)	—
Breslau	114	Darwin (1871)	—
Prussia	113	Darwin (1871)	—
Posen	108·35	Von Bergmann (1883)	1819–1873
Prussia	107·64	Düsing (1884)	1875–1887
Prussia	107·20	Von Fircks (1898)	1820–1867
Livonia	105	Pearl (1913)	—

(c) The secondary sex-ratio varies among different races living under the same topographical and climatic conditions.

Table X.

Sex-ratio of White and Coloured Races.

Locality	Authority	Ratio for Whites	Ratio for Coloured
U.S.A.	Jastrzebski (1919)	105·7	100·0
Cape Colony	„	105·4	102·6
Columbia	„	105·0	100·0
New York	„	104·5	101·6
New Orleans	„	102·0	98·2
U.S.A. (1st births)	Little (1920)	115·51	93·61
Columbia	Nichols (1907)	106·2	103·0
Cuba	Heape (1909)	108·42	101·2

It will be noted that everywhere and in every case the ratio for coloured races is lower than that for whites and this suggests that the difference is significant.

(d) There are reasons for holding that the production of a profoundly unusual sex-ratio is characteristic of a particular individual. Instances, not a few, are known in which a male of some monotocous species has produced in different matings none but female or none but male offspring. But the monotocous female, producing usually but one offspring at a time, is not a suitable material with which to test this hypothesis. An appeal must be made to the polytocous rodent. King (1918) has found that in the case of the albino rat it was possible, starting with two pairs of rats from the same litter, to found two strains, one of which produced a high proportion of males, the other a preponderance of females. The progeny of one pair (pair A) were bred brother to sister without selection for six generations in order to build up a homozygous and uniform race. After this, selection was practised, the brothers and sisters being chosen from litters which showed a preponderance of males. In line B the selection after the sixth generation was made from litters showing a preponderance of females. After fifteen generations of such inbreeding and selection, the sex-ratio at birth in line A was 125 : 100, in line B 83 : 100. The habitual production of an unusual sex-ratio can be the expression of a genetic constitution. It is possible to breed for a preponderance of one sex.

3. The secondary sex-ratio varies from year to year, but the amount of variation is small. Parkes (1926) has pointed out that in the case of the records for England and Wales the variation is not around equality but about a point above equality, the mean of the extremes being 103 : 75. There are thus two forces at work, one that causes a constant preponderance of males, and another which causes the excess of males to vary in amount from year to year (see Gini, 1908; Düsing, 1884).

4. The secondary sex-ratio varies with the season of the year. It is not improbable that the breeding season is one in which the general physiological condition of the individual is above the average, and it is of interest therefore to compare the secondary sex-ratio following conception during the breeding season with that following conception at other times. The human birth-rate actually shows a slight variation in spite of the fact that nearly all traces of the primitive breeding season have been obliterated by social habits. It is found that the sex-ratio is low for births resulting from conceptions at the seasons of greatest fertility and high at the times of lowest birth-rate.

Dighton (1922) gives the following figures for the greyhound:

Table XI.

	Number of puppies	Dogs	Bitches	Secondary sex-ratio
January	726	382	344	111 : 100
February	593	318	275	112 : 100
March	906	468	438	106 : 100
April	794	408	386	105 : 100
May	1129	587	542	108 : 100
June	892	486	406	110 : 100
July	932	480	452	106 : 100
August	669	354	315	112 : 100
September	314	173	141	122 : 100
October	250	122	128	90·6 : 100
November	209	107	102	104 : 100
December	122	57	65	87·6 : 100

Heape (1908) and Wilckens (1886) supply the following data:

Table XII.

		Secondary sex-ratio		
Animal	Numbers	Whole year	Warm months	Cold months
Horse	16091	97·9 : 100	96·6	97·3
Cattle	4900	107·3 : 100	114·1	103·0
Sheep	6751	97·4 : 100	102·1	94·0
Pig	2357	111·8 : 100	115·0	109·3
Dog	17838	118·5 : 100	126·3	122·1

It should be mentioned that Bonnier (1923) maintains that the Swedish vital statistics do not confirm Heape's conclusions.

Parkes (1924) gives the following figures for the albino mouse:

Table XIII.

	Numbers	Males	Females	Secondary sex-ratio	Percentage of males
March	25	12	13	100 : 100	
April	82	42	40	105·0 : 100	49·7 ± 1·78
May	90	44	46	95·7 : 100	
June	157	78	79	98·8 : 100	
July	163	88	75	117·4 : 100	
August	231	138	93	148·5 : 100	56·5 ± 1·27
September	187	100	87	115·0 : 100	
October	96	56	40	140·0 : 100	
	1031	558	473	118·0 : 100	54·2 ± 1·04

King and Stotsenburg's (1915) figures for the rat also support the contention that the season of the highest birth-rate is that of the lowest proportion of males. Sumner, McDaniel, and Huestis (1922) suspect a seasonal variation also in the case of the deer-mouse, *Peromyscus*, which however gives the highest sex-ratio in the season opposite to King's.

5. The secondary sex-ratio varies with different matings and the disturbance seems to be related to the relative physiological condition of the parents at the time. In the experience of many poultry breeders, the first lot of eggs laid by a pullet yields a preponderance of male chickens, whereas as the season advances and

the pullet ages, the proportion of males steadily decreases. Jull (1923), using a sex-linked cross in order to preclude errors, recorded the sex of the chickens hatched from eggs of 45 hens during their first year of production. The observations were repeated for three years and the secondary sex-ratio was found to be 48·41 expressed as a percentage. Analysis of the figures gave the following table:

Table XIV[1].

Eggs			Secondary sex-ratio
o	to	20	62·91 ± 1·44
11	,,	40	57·46
41	,,	60	45·00
61	,,	80	44·61
81	,,	100	37·65
101	,,	120	32·53 ± 1·15

6. The secondary sex-ratio is profoundly disturbed as a result of interspecific and intervarietal crosses. Haldane (1922) has pointed out that in any such cross the sex that is absent, rare, or sterile is the heterogametic sex.

Table XV.

INSECTS. (In Lepidoptera the female is digametic; in *Drosophila*, the male)

Mother	Father	Males	Females	Author
Nyssia graecaria	*Lycia hirtaria*	65	o	Harrison (1916)
N. zonaria	,, ,,	208	o	,, (1919)
,,	*Paecilopsis isabellae*	32	o	,,
,,	,, *pomonaria*	90	o	,,
,,	,, ,, (inbred)	71	7	,,
,,	,, *lapponaria*	93	o	,,
,,	,, ,, (inbred)	62	3	,,
Lycia hirtaria	,, *pomonaria*	86	75	,, (1916)
,, ,,	,, ,,	190	14	,,
Poecilopsis isabellae	*Lycia hirtaria*	38	32	,,
P. lapponaria	*Poecilopsis pomonaria*	38	39	,, (1917)
Oporabia dilutata	*Oporabia autumnata*	6	o	,, (1920a)
Tephrosia bistortata	*Tephrosia crepuscularis*	378	12	,, (1920b)
Drosophila melanogaster	*Drosophila melanogaster*	—	—	Lynch (1919)
(fused)	(normal)	o	823	,,
(fused XXY)	,,	9	744	,,
(rudimentary)	,,	10	923	,,
(rudimentary XXY)	,,	93	649	,,
Drosophila melanogaster	*Drosophila simulans*	2	3552	Sturtevant (1920)

[1] But see Lambert and Knox (1926).

Table XV (contd).

BIRDS (female digametic)

Mother	Father	Males	Females	Author
Turtur orientalis	*Columba livia*	13	1	Whitman and Riddle (1919)
Streptopelia risoria	,, ,,	38	0	,,
S. alba-risoria	,, ,,	11	0	,,
S. risoria	*Zenaidura carolinensis*	16	0	,,
Gallus domesticus	*Phasianus colchicus*	100	—	Lewis Jones (quoted by Haldane)
Phasianus reevesi	,, *torquatus*	{161	6}	Smith and
,,	,, *versicolor*			Thomas (1913)
Tetrao urogallus	*Tetrao tetrix*	40	8	Suchetet (1897)
Gallus domesticus	*Pavo nigripennis*	2	0	Trouessart (1907)

MAMMALS (male digametic)

Mother	Father	Males	Females	Author
Bos taurus	*Bison americanus*	6	39	Boyd (1914)
,,	*B. bonasus*	1	3	Ivanov (1913)

The effects of hybridisation upon the sex-ratio is a subject of considerable anthropological interest. Three types of hybridisation are possible: (1) between white races, (2) between coloured races, and (3) between white and coloured races. As a result of an investigation of the effect of hybridisation, Little (1920) concluded that a higher preponderance of males resulted from hybrid white matings than from pure white matings, and that hybrid coloured matings gave a higher preponderance of females than did pure coloured matings. This is in agreement with Lewis (1906) who found that unions of Spanish, Italian, and French male emigrants with native-born Argentine females produced a higher secondary sex-ratio than pure Argentine alliances or pure alliances of any of these nationalities in Buenos Ayres; and also that unions of Argentine males with females of foreign nationalities gave a higher sex-ratio than pure Argentine matings. Pearl and Pearl (1908) came to the same conclusion. Powers (1877) states that, in the case of black and white hybrids, there is a large excess of girls among half-breeds in California, and Kohl (1859) notes that in the northern parts of the United States females preponderate in the progeny of French men with Indian women. Starkweather (1883) found in the mulatto a 12 to 15 per cent. excess of females, while in the whole population males were in excess. An excess of females is reported by Görtz (1853) among the offspring of Dutch men and Malay women in Java, and this has been confirmed by Waitz (1859). Jastrzebski (1919) states that for the years 1910–15 the ratios of males to 100 females in New York were: whites, 104·0; negroes, 99·9; and mulattoes, 97·9. Bugnion (1910) states that in a hybrid race formed by settlers in a colony of negroes there was a marked preponderance of females. It seems therefore possible to conclude (1) that crosses between white races produce an excess of males over pure white matings; (2) that hybridisation between coloured races produces an excess of females above pure coloured matings; and (3) that hybrids of white and coloured races show an excess of females above the pure matings of either race.

7. The secondary sex-ratio varies with the parity, *i.e.* with the chronological number of the pregnancy.

In the case of the human, the dog, and the mouse, it has many times been noted that there is a continuous drop in the sex-ratio at each succeeding pregnancy (Wilckens (1886), Punnett (1903), Bidder (1878), Copeman and Parsons (1904)). King and Stotsenburg (1915) found that the same rule obtained in the case of the rat:

Table XVI.

Sequence of litters	No. of litters in the series	Individuals	Males	Females	Secondary sex-ratio
1	21	131	72	59	122·0
2	21	162	85	77	110·4
3	18	127	64	63	101·6
4	15	96	41	55	103·1

Parkes (1924), working with mice, gives the following table:

Table XVII.

	Total	Males	Females	Secondary sex-ratio	Percentage of males
First births	242	134	108	124·2	55·4 ± 2·14
Subsequent births	190	114	76	150·0	60·0 ± 2·39

Parkes points out, however, that most of the second and higher births occurred at the end of the breeding season when the sex-ratio is at its highest, and that too much must not be inferred from his figures until the experiment has been repeated.

8. It has been suggested that the sex-ratio varies with the time relation of successive conceptions. Rumley Dawson (1921), for instance, maintained that in the case of the monotocous female the right ovary elaborated only male-producing ova, the left only female-producing, and that the ovaries function alternately, ovulation occurring in one ovary at one oestrous period, in the other at the next. Knowing then the sex of the first offspring of the female, and keeping a record of all oestrous periods, including those suppressed by the first pregnancy, it is possible, according to this theory, to arrange services so that subsequent offspring produced by this female shall be of the selected sex and likewise to foretell their sex. In order that an offspring of the same sex as the first shall be produced, all that is necessary is to see that conception shall coincide with an oestrous period which, when referred back to the one associated with the conception of the first offspring, is an odd number, 9th, 11th, 13th, and so on. If progeny of the opposite sex are desired, then conception must be made to coincide with oestrous periods with even numbers, 10th, 12th, and so on. It should be stated that a considerable number of experienced stock-breeders in this and in other countries claim that their observations entirely support this theory. Nevertheless it cannot be brought into harmony with established scientific facts, and therefore cannot be accepted on its present-day evidence. The theory is supported by a collection of selected statistical data applied without proper statistical treatment, and those cases which do not fit into the scheme are airily dismissed, whilst the great body of established facts which supports other theories and cannot support this particular one is neglected. Variations of this theory are numerous, and like it are based upon the conception that in the human, horse, and cattle, the female is digametic. Many believe that offspring of either sex can be obtained at will by persuading the semen of the male to flow to the right or the left of the body of the female, towards the left ovary or to the right, that is. This is ensured by the female lying on one side or the other after coitus, or standing on a slope. The matter has been tested experimentally and found wanting. Doncaster and Marshall (1910) have shown that unilateral ovariotomy in the rat does not result in the production of offspring of one sex only, and the cogency of these

experimental results cannot be dismissed by the statement that it is too far a cry from the rat to the human female. If the breeder really desires to have this theory tested, the way is simple, for it can readily be shown that unilateral ovariotomy in the horse or in the cow is not followed by the production of offspring of one sex only, and that the production of both male and female progeny is not to be explained by any regeneration of the imperfectly removed ovary. As it is, the believer in this theory may well be content, for a calf must be either a male or a female; some matings undoubtedly produce a preponderance of female calves, and in any case if a calf of the wrong sex appears the breeder can always find some satisfying explanation for the unexpected.

9. In the case of the rabbit it has been shown that the sex-ratio is related to the chronological order of the service of the buck; in the first service group there is a preponderance of males and then an increasing preponderance of females as the number of services increase. Hays (1921) obtained the following results (see Düsing's figures for the horse, Table V):

Table XVIII.

Service	...	1st	5th	10th	15th	20th
Sex-ratio	...	56·33	43·50	44·44	54·64	21·87

10. Statistical evidence has been presented, sometimes supporting, at other times contradicting, the suggestion that the sex-ratio is affected by the relative ages of the parents, the offspring being mostly of the same sex (or of the opposite sex) as the older (or as the younger) parent. Hofacker (1828) and Sadler's (1830) law —that the sex of the offspring is that of the older parent—finds no support in the result of critical inquiry: it is contradicted by the work of Schultze (1903) on mice, for example. According to some data, the age of the mother has a relation to the sex of the offspring, younger mothers producing a preponderance of males (or of females). It must not be forgotten that in the majority of cases the data upon which these theories are based have not been collected by biometrical experts, nor can their genuineness be absolutely guaranteed. Other data seem to suggest that the sex of the offspring tends to be that of the more (or of the less) vigorous parent. For example, the theory of Starkweather (1883) suggested

that the "superior" parent tended to beget offspring of the opposite sex, but since it is impossible as yet to define "vigour" and "superiority" in accurate physiological terms, such theories are not suitable for scientific discussion. There is no experimental basis for such conceptions.

11. It has been suggested that the sex-ratio varies with the time of service during the oestrous period. Pearl and Parshley (1913), in testing this theory first propounded by Thury (1863) and later by Düsing (1884) collected data from stock-breeders, and found that out of a total of 480 calves, 255 were male and 225 female, and that the sex-ratio among calves resulting from service during the middle period of the heat was 115·5, and in the case of service late in heat, 154·8. Later data collected by Pearl (1917) did not support the suggestion that service early in heat resulted in a female calf, late in heat, in a male.

The suggestion that there is a relation between the sex-ratio and the size of the litter in polytocous animals is not supported by the results of Wentworth (1914) in the case of dogs and pigs, of King and Stotsenburg (1915) in the case of rats, of Parker and Bullard (1913), Machens (1915) and also of Parkes (1923) in the case of pigs.

Two main facts emerge from a consideration of these data: (1) the secondary sex-ratio is not equality, and (2) it is distinctly variable within limits in a given species.

Reasons for holding that the primary sex-ratio is not equality have been given. It is necessary to find out whether or not the secondary sex-ratio is identical with the primary or whether a sexually selective prenatal mortality so operates as to make these ratios markedly dissimilar.

Parkes (1923) has shown that prenatal mortality in mice falls preponderatingly upon the male[1]. This is also the case in pigs, for Parkes (1925) examined some 500 pig foetuses of different sizes and assuming that the size of the foetus could be regarded as an indication of its age, divided them into four groups, largest, smallest, and two intermediate classes. The following table shows that in this series the foetal sex-ratio varied with the age-group

[1] In this connection see paper by MacDowell and Lord, 1926.

in such a way that the largest (= oldest) group exhibited the lowest ratio, the smallest (= youngest) the highest, whilst the other two classes fitted into this ascending series. An exactly similar result was obtained by Crew (1925) working also with pig foetuses.

Table XIX.

Average weight of foetuses in uteri

	0–100 gm.			101–300 gm.			301 + gm.			Total		Grand total
	♂	♀	Total	♂	♀	Total	♂	♀	Total	♂	♀	
Complete uteri	69	57	126	20	13	33	46	41	87	135	111	246
Incomplete uteri	97	58	155	45	36	81	54	47	101	196	141	337
Total	166	115	281	65	49	114	100	88	188	331	252	583
Sex-ratio	144·5			134·0			113·5			131·4		
Percentage males	59·1			57·0			53·2			56·8		
Probable error of percentage	±1·98			±3·12			±2·45			±1·38		

In the case of the human subject ample data which show that there is a high prenatal mortality and that this is sexually selective are available. The secondary sex-ratio in the human as recorded by different authorities is between 90 to 110 males per 100 females. The prenatal death of a zygote is followed by autolysis and re-absorption in the polytocous mammal and this occurs also in the case of an early embryo in the monotocous female. Older foetuses,

Table XX.

Amount of Abortion in Man.

		Percentage	
		Births	All pregnancies
Priestley (1887)	...	30·0	23·0
Routh (1914)	...	27·4	20·0
Parkes (1924)	...	19·8	—
Auerbach (1912)	...	10·6	—
Taussig (1910)	...	—	30·3
Pearson (1908)	...	—	28·5
Rauber (1900)	...	—	24·0
Malins (1903)	...	—	19·3
Williams (1923)	...	—	16·2
Franz (1898)	...	—	15·4
Whitehead (1914)	...	16·0	14·0

dying, are extruded as abortuses. It is very probable that even were it possible to compute the entire amount of abortion, the full figures for prenatal mortality would not be available. Because of this, and because mortality does not necessarily result in abortion, it follows that any figure which is given for prenatal mortality must fall short of the true amount. However, it will at least serve as a rough guide. The above table (XX) summarises the considerable number of estimates which have been made.

It will be seen from the above table that even the amount of mortality which it has been possible to compute is very considerable and is quite sufficient to influence the sex-ratio if the sex-incidence is at all peculiar. As regards this point, the following table gives the various estimates which have been made of the sex-ratio of abortions.

Table XXI.

Sex-ratio of Abortions.

Authority	Age of abortion	Number	Sex-ratio
Körösy (1898)	—	3781	152·4
Rauber (1900)	—	—	159·0
Rust (1902)	1st 6 months	454	101·8
Lenhossék (1903)	3–6 months	156	160·0
Carvallo (1912)	Up to 4th month	—	250·0
Pinard and Magnan (1913) ...	—	1229	101·1

The above figures are the ratios for the aggregate prenatal mortality, but Auerbach (1912) shows in the following table that the excess of males in prenatal mortality increases as conception is approached.

Table XXII.

Sex-ratio of Abortions according to Age.

Time	Numbers	Males	Females	Sex-ratio
4th month	1333	928	405	229
5th month	1041	645	396	163
6th month	943	506	437	116
7th month	750	402	348	116

A similar fact has been shown as regards the gross amount of mortality. Günther (1923) has calculated that the curves for both the amounts and sex-incidence of prenatal mortality follow a logarithmic formula back into gestation, which means that the early wastage of males must be very considerable and more than it is possible to calculate. However, certain authors have endeavoured to calculate the primary sex-ratio by using the figures available. The following table gives their conclusions.

Table XXIII.

Estimates of Sex-ratio at Conception.

Authority			Ratio
Auerbach (1912)	...	...	116·4
Lenhossék (1903)	...	...	111·0
Dawson (1921)	...	...	110·0
Jendrassik (1911)	...	...	108·7
Bernouli (1841)	...	...	108·2

Though these estimations are probably of no great accuracy, it is safe to maintain that the primary sex-ratio is very much higher than the secondary. This is in agreement with the results of examination of pig, cattle, and mouse foetuses.

The amount and sex-incidence of prenatal mortality find direct continuation in the nature of still-births. The question of still-births is, however, complicated by the fact that two factors play a part in determining their frequency. Firstly, there are the still-births which are dead when parturition sets in, and secondly, there is the class which is caused purely by difficult labour (dystocia). However, it is possible to make some differentiation between these two. If still-births were mainly due to dystocia they would be most frequent when dystocia is most common, *i.e.* among first-births. Parkes (1924) has shown that this is not the case and that still-births, like abortions, occur predominantly at later pregnancies. Hence still-births are presumably not mostly due to dystocia; the majority must be dead before parturition occurs. If this is the case, one would expect still-births to be due to the same causes as abortions and therefore to present similar features. In the first place the percentage of still-births is very considerable, and, as in the case of abortions, the sex-incidence is unequal and falls largely upon the males. The following table sums up the large number of estimations of the sex-ratio of still-births.

THE SEX-RATIO 167

Table XXIV.

Sex-ratio of Still-Births.

Authority	Country	Sex-ratio
Bodio (1894) ...	France	142·2
	Norway	124·6
	Holland	127·1
	Germany	128·3
	Hungary	130·0
	Italy	131·1
	Belgium	132·0
	Denmark	132·0
	Austria	132·1
	Sweden	135·0
	Switzerland	135·0
Davis (1918) ...	United States	137·1
Dawson (1921) ...	—	138·0
Düsing (1884) ...	Prussia	129·09
Heape (1909) ...	Cuba	144·45
Hirsch (1913) ...	Germany	127·9
King (1921) ...	America	131·06
Lewis (1906) ...	Europe	120–170
Nichols (1907) ...	—	131·6
Quetelet (1872) ...	—	133·5

These ratios all centre very closely round 135, which figure, though below the ratio for abortions, is considerably above the ratio for live births. The greater mortality of males is maintained after birth. The following table gives the ratio of mortality by three monthly periods from the time of conception to the end of the first year of life. It will be noticed that the ratio shows a continuous decline.

Table XXV.

Sex-ratio of Mortality from Conception to end of 1st Year of Life.

	Time	Sex-ratio of mortality	Authority
Prenatal ...	4–7 months	156·5	Auerbach (1912)
	7–9 ,,	138	(Average figure
Postnatal ...	0–3 ,,	131·5	Registrar-General)
	3–6 ,,	121	,,
	6–9 ,,	120	,,
	9–12 ,,	113	,,

Parkes (1926) has shown that when this table is graphed it is seen that the curve for the decline in the sex-ratio of mortality is very nearly linear and it is interesting to project back this curve to the time of conception, and thus to arrive at a presumable ratio for the sex-ratio of mortality during the first three months of gestation, a time for which no actual figure can be observed.

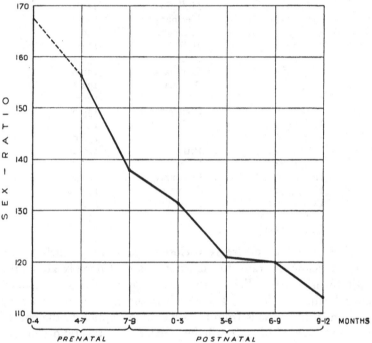

Fig. 32. The primary sex-ratio in man. (*After* Parkes.)

On the basis of this graph it would seem that the sex-ratio of mortality for the first three months of gestation is probably 166–168.

The Causes of this Prenatal Mortality.

It is seen, then, that there exists a very considerable prenatal sexually selective mortality. Far more males than females are conceived in those forms *in which the male is the digametic sex*, and far more males than females perish prenatally so that the secondary sex-ratio approaches equality.

In certain instances it can be shown that the disturbance of the sex-ratio is due to the action of an unbalanced sex-linked lethal factor. Prenatal mortality due to the action of an unbalanced sex-linked lethal factor falls upon the heterogametic sex. When more than one sex-linked lethal are present in the genotype, the sex-ratio will depend on the linkage value between them. For example, in *Drosophila melanogaster*, lethal 1 (l_1) is located at 0·7, lethal 3 (l_3) at 26·5 on the X-chromosome, and therefore about 25 per cent. of crossing-over will occur. The female with each of these lethals in the simplex state will have the constitution $(l_1 L_3 X) (L_1 l_3 X)$ and will elaborate four sorts of eggs in the following proportions:

$(l_1 L_3 X)$	$(L_1 l_3 X)$	$(L_1 L_3 X)$	$(l_1 l_3 X)$
3	3	1	1

and when these are exposed to fertilisation by the sperms elaborated by a wild-type male $(L_1 L_3 X) Y$ the following zygotes will result:

$(l_1 L_3 X)(L_1 L_3 X)$	$(L_1 l_3 X)(L_1 L_3 X)$	$(L_1 L_3 X)(L_1 L_3 X)$	$(l_1 l_3 X)(L_1 L_3 X)$
3	3	1	1
$(l_1 L_3 X) Y$	$(L_1 l_3 X) Y$	$(L_1 L_3 X) Y$	$(l_1 l_3 X) Y$
3	3	1	1
(die)	(die)	(lives)	(dies)

so that only one male in every eight will live and a sex-ratio of 12·5 : 100 will result. It is probable that sex-linked lethals and semi-lethals (which allow a few of the XY individuals to come through) are at the bottom of most of the excess of male foetal mortality. Jewell (1921) found that the foetal sex-ratio in cattle was 123 : 100, so that there must be a selective elimination of male foetuses. The specific action of a lethal is not to be demonstrated if that action is exerted during the earliest stages of embryonic development; it has to be assumed that the physiological derangement or the ·anatomical abnormality was such that, becoming expressed in the earliest stages of development, it rendered the zygote incapable of pursuing the further developmental stages. In the case of the bison × cattle cross it has been shown that the lethal effect is such as to lead to severe dystocia preventing the birth of the male calf. The abnormal sex-ratio following interspecific crosses may be due to some such cause as the action of a sex-linked lethal complex.

In the case of *Drosophila melanogaster*, it has been shown also that primary and secondary non-disjunction lead to a profoundly unusual sex-ratio, and in these cases there is some definite abnormality in the chromatin content and in the genotype of the individuals that succumb.

Selective mortality of the developing zygotes may be based upon a differential susceptibility to disease. Federley (1911), for example, obtained in two successive generations of *Pygaera pigra*, the chocolate-tip moth, none but females, and found that the lack of males was due to an inherited disease which had the effect of making the blood of the male larvae abnormal, so that they were killed off in the larval stage. The females were unaffected but transmitted the disease to their male offspring. In such a case as this the action of a sex-limited lethal characterisation demonstrates a physiological distinction between the sexes. Goldschmidt (1923) was able to show that in the case of the caterpillars of the gipsy moth attacked by Flacherie, a disease which affects the caterpillar about the time of the fifth moult, it did not affect the males, as these were already in the pupal stage and thus secure from infection, but that it did affect the females, causing among them a high death-rate, and resulting in a profoundly disturbed sex-ratio. It is seen, then, that differences in the life history of the two sexes may provide an opportunity for the partial or complete elimination of one.

This selective elimination is also affected by favourable and unfavourable conditions during pregnancy. Parkes (1923) has shown that mice which are allowed to become pregnant immediately after parturition and whilst still suckling their young produce smaller litters in which the proportion of females is markedly raised. Corpora lutea counts showed that in such cases the number of foetuses was not unusual, and that the smaller litter-size was due to an intensified selective elimination of the males. The same explanation can be applied to the fact that the secondary sex-ratio among illegitimate children is lower than that among legitimate—the lack of prenatal care and hygiene resulting in an intensification of the forces that always make it relatively difficult for the mammal to beget male offspring and for the bird to produce as many females as males. The secondary sex-ratio is

highest among those peoples and those herds in which the highest
degree of prenatal hygiene is practised. It is probable that amongst
the coloured people there is a higher prenatal mortality and that
this is the case also among the hybrids of coloured races and of
white and coloured, whereas the mortality of hybrids of two white
races is relatively very low in consequence of the hybrid vigour
or heterosis that results from such a mating.

An unusual sex-ratio may be the reflection of complete sex-
reversal. It is an established fact that an individual with the chromosome constitution of one sex, a genotypic male or female, as
the case may be, may come to possess the functional sex-equipment
and capacity of the opposite. A hen, XY in genotypic constitution,
may function as a male, and mated with a hen will produce offspring, the sex-ratio amongst which will be 50 : 100. If, on the
other hand, a female amphibian or mammal, XX in genotypic
constitution, functions as a male, its progeny will consist solely
of females.

(a)	XY	×	XY		(b)	XX	×	XX
	X : Y		X : Y			X		X
	XX	XY XY	YY				XX	
	1 ♂	2 ♀♀	dies				♀	

Goldschmidt (1920, 1923), Harrison (1919) and others have
obtained species hybrid broods that were largely or entirely of
one sex. Harrison was able to show that in one of his cases the mortality was not sufficiently high to account for the results obtained;
that it was not a case of a sexually selective prenatal mortality, and
both Goldschmidt and Harrison have presented evidence which
strongly indicates that the results are due to a sex-transformation
of half the individuals concerned. Unisexual broods in these forms
are to be regarded as the final stage of intersexuality. Sturtevant
(1920) records similar unisexual broods in *Drosophila melanogaster*
× *D. simulans* (Diptera) crosses, but finds that the cause of this
is a selective mortality.

C. THE TERTIARY SEX-RATIO

The tertiary sex-ratio will be identical with the secondary unless
during the period birth–maturity a sexually selective mortality is

operating. Any difference between the secondary and tertiary ratios is simply the result of a selective postnatal mortality and serves as an indication of the relative postnatal survival value of the sexes.

The only material available for examination is the human. The returns of the Registrar-General for 1913 reveal the following facts. In the age groups 0–5 years the sex-ratio of infantile mortality is 113·4; in the 5–10 group it is 100·7; in the 10–15 group it is 93·3. This period, 10–15, is, in fact, the only one during which more females than males die, and this is held to be the result of the exhaustion of puberty in the female and the incidence of tuberculosis (Schultz, 1918, and Stewart, 1910–1911). From 15 years on the sex-ratio of mortality rises steadily to 50, when a slight decline sets in. This excess of male mortality is probably mainly due to occupational stress. The effect of this selective mortality is seen in the swing of the sex-ratio of the population; 104 : 100 at birth, it is reduced to 102 : 100 at the end of the first year, and to 101·5 : 100 at the end of the second. In the third to fifth years it falls to 101·3 : 100. From 5–10 it becomes 99·9 : 100; from 10–15 it becomes 94·2 : 100. During the 15–20 age period there is a rise following upon the increased mortality of females during the 10–15 age group. From this point on there is a continuous drop, with the exception of a slight rise between 40 and 50 during which period there is an increased mortality of females from reproductive disorders during the period 40–45 years. In old age (85 years) the sex-ratio is only 55·2 : 100.

It is impossible as yet to define the causes of this postnatal sexually selective mortality. The action of semi-lethal factors, differences in occupational risks, and such like, cannot explain all the facts, and all that can be said at present is that for reasons as yet unknown the male exhibits an inherently inferior resistance to the stresses of the act of living.

D. PARTHENOGENESIS AND THE SEX-RATIO

In the case of a species in which, in addition to the sexual mode of reproduction, parthenogenesis also obtains, it is manifest that any environmental agency affecting pairing must affect the sex-ratio, having a direct and cumulative effect on this. The end result will differ according to whether parthenogenetic eggs yield males or females.

Williams (1917), in describing the results of certain breeding experiments on the white fly, *Aleurodes vaporariorum*, discusses the question of the sex-ratio and parthenogenesis and points out that the change in the sex-ratio can be treated mathematically if it is assumed (1) that the size of the colony remains constant; (2) that there is no differential mortality between the sexes or between the fertilised and unfertilised eggs; (3) that the fertilised eggs produce equal numbers of each sex (as has been found in *Aleurodes vaporariorum*); and (4) that the species is monogamous.

If in a normal locality all the individuals pair as far as possible, whatever be the result of parthenogenetic eggs, there will be a permanent position of stability at 50 per cent. of each sex, and if some of the females are accidentally destroyed in one generation there will be a return to equality in the next generation. If, however, the number of males is reduced below 50 per cent. in a monogamous species, not all the females will be able to pair. The result will now depend on the sex produced by the unfertilised eggs. If males, there will be an increase in the percentage of males in the next generation followed by a return in following generations to equality. But if females are produced, the original slight accidental reduction of males will be further increased in each successive generation, owing to those females which have not paired producing families of females only as large as those of both sexes produced by the paired females. Considering a colony of 200 individuals of which normally 100 are of each sex, then should the number of males be reduced to 98, the resulting number of males in the successive generations is as follows:

$$
\begin{array}{lll}
P_1 = 98 & F_4 = 75 & F_9 = 9 \\
F_1 = 96 & F_5 = 60 & F_{10} = 5 \\
F_2 = 92 & F_6 = 43 & F_{11} = 2 \\
F_3 = 86 & F_7 = 28 & F_{12} = 1 \\
& F_8 = 16 &
\end{array}
$$

It is seen that without any further accidental loss or indirect influence the number of males will be reduced in twelve generations from 49 per cent. to less than 1 per cent. The result is seen graphically in Fig. 33 which shows that the change is at first slow,

becoming more rapid in the region of 25 per cent. males and finally
slowing off again, so that in a large colony a few males might
persist for many generations.

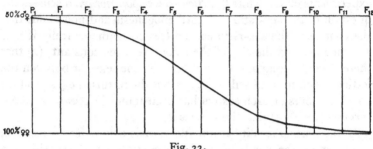

Fig. 33.

The above result only holds for monogamous species. Polygamous species will return at once to the equilibrium position of
equality of sexes so long as the males left in the colony are sufficient
to pair with all the females. When they are unable to do this, then
a process similar to the above will start, but the change will be less
rapid.

The effect of failure to pair on the normal 50 per cent. equilibrium
will be somewhat different, according to whether it is the males or
the females which are affected, and it also depends as before on
the sex produced by the parthenogenetic eggs. The four combinations are treated below, and each is, for completeness, tabulated
for three different strengths of disability, namely, 10 per cent.,

Table XXVI.

Males affected by conditions.

A. *"Female-producing" race.*

Colony consisting originally of 100 males and 100 females.

		10 per cent.		50 per cent.		100 per cent.	
		♀♀	♂♂	♀♀	♂♂	♀♀	♂♂
P_1	...	100	100	100	100	100	100
F_1	...	110	90	150	50	200	—
F_2	...	119	81	175	25	200	—
F_3	...	127	73	187	13	200	—
F_4	...	134	66	194	6	200	—

50 per cent., and 100 per cent., that is, in the first case ten out of
every hundred of the sex specified do not pair, in the second case
fifty, and in the last case none.

The three cases are shown graphically in Fig. 34.

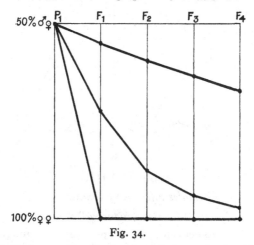

Fig. 34.

In the second case the elimination of the males is similar to the
first but more rapid. In the third, all the males are lost in the first
generation. If the normal conditions return after a few generations,
then the curve will change to the form found in Fig. 33 unless
the species is polygamous, in which case there will be a return
to equality.

Table XXVII.

B. "Male-producing" race.

		10 per cent.		50 per cent.		100 per cent.	
		♀♀	♂♂	♀♀	♂♂	♀♀	♂♂
P_1	...	100	100	100	100	100	100
F_1	...	90	110	50	150	Nil	200
F_2	...	100	100	100	100	Nil	Nil
F_3	...	90	110	50	150	—	—
F_4	...	100	100	100	100	—	—

The above is shown graphically in Fig. 35.

In this case there is an alternation between equality of sexes
and an increased number of males except in the extreme form where

the males increase to 100 per cent. in the first generation and as a result the colony dies out. The result would be the same whether the species was polygamous or monogamous.

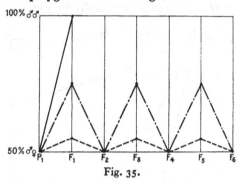

Fig. 35.

Table XXVIII.

Females affected by conditions.

A. *"Female-producing" race.*

		10 per cent.		50 per cent.		100 per cent.	
		♀♀	♂♂	♀♀	♂♂	♀♀	♂♂
P_1	...	100	100	100	100	100	100
F_1	...	110	90	150	50	200	Nil
F_2	...	118	82	167	33	200	—
F_3	...	131	69	180	20	200	—
F_4	...	147	53	189	11	200	—
F_5	...	164	36	195	5	—	—
F_6	...	178	22	197	3	—	—

The above is shown graphically in Fig. 36.

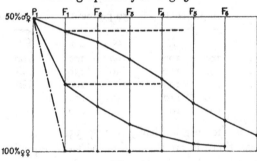

Fig. 36.

In these cases only the change of percentage in the first generation is due to the effect of pairing; after this time, as there are many more females than males, a few not able to pair will make no difference to the result, which then follows the normal curve shown in Fig. 33. If the species is polygamous, all the generations after the first will have the same percentage of the two sexes as shown by the dotted lines in Fig. 36.

Table XXIX.

B. "*Male-producing*" race.

	10 per cent.		50 per cent.		100 per cent.	
	♀♀	♂♂	♀♀	♂♂	♀♀	♂♂
P_1 ...	100	100	100	100	100	100
F_1 ...	90	110	50	150	Nil	200
F_2 ...	90	110	50	150	Nil	Nil
F_3 ...	90	110	50	150	—	—
F_4 ...	90	110	50	150	—	—

The above is shown graphically in Fig. 37.

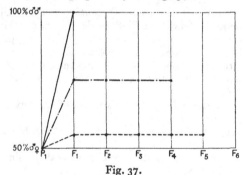

Fig. 37.

In this case the percentage of males is permanently raised. In the extreme form in which none of the females pair, the first generation consists entirely of males and the colony dies out. Polygamy will have no effect on the result. If the conditions affecting pairing be removed, the colony will return in the next generation to the stable position with an equal number of each sex.

It will be seen from an inspection of the above results that in the "female-producing" races there is a constant liability to loss or reduction of the male sex, while in the "male-producing" races the tendency is in the opposite direction. It is to be noted that the relative length of life of the two sexes will affect the apparent sex-ratio obtained by a random collection without, however, affecting the proportions originally emerging from eggs or pupae. As the adult females live much longer than the males the proportion of the former will be too high in counts obtained from random collections.

BIBLIOGRAPHY (D)

ADLER, L. (1917). *Loc. cit.* (B).

AUERBACH, E. (1912). Das wahre Geschlechtsverhältniss des Menschen. *Arch. Rass. Ges. Biol.* 9, 10–17.

BALTZER, F. (1914). *Loc. cit.* (B).

BÄLZ, VON, E. (1911). Das Verhältnisszahl der Geburten in verschiedenen Ländern. *Korresp.bl. deut. anthrop. Ges.* 42.

BERGMANN, E. (1883). *Zur Geschichte der Entwicklung deutscher, polnischer und judischer Bevölkerung der Provinz Posen seit 1829.* Tübingen.

BERNOULI, C. (1841). *Handbuch der Populationsstatistik.* Ulm.

BIDDER, F. (1878). Ueber den Einfluss des Alters der Mutter auf das Geschlecht des Kindes. *Zeit. Geburtsh. Gynäk.* 2, pp. 358–364.

BODIO, L. (1894). *Movimento della popolazione.* Rome.

BONNIER, G. (1923). Ueber die Realisierung verschiedener Geschlechtsverhältnisse bei *Drosophila melanogaster. Zeit. indukt. Abst.* 30, pp. 283–286.

—— (1923). On Alleged Seasonal Variation of the Sex Ratio in Man. *Ibid.* 32, pp. 97–107.

BOYD, M. M. (1914). Crossing Bison × Cattle. *Jour. Hered.* 5, pp. 187–197.

BUGNION, E. (1910). Les cellules sexuelles et la ·détermination du sexe. *Bull. Soc. Vaud. Sci. Nat.* 46, pp. 263–316.

CAMPBELL, J. (1870). On Polygamy, its Influence in Determining the Sex of our Race, and its Effect on the Growth of Population. *Jour. Anthrop.* 8, p. cviii.

CARVALLO, E. (1912). La masculinité dans les naissances humaines. *C. R. Acad. Franç. Avan. Sci.* 41, pp. 145–146.

CHAMPY, C. (1921). *Loc. cit.* (B).

COLE, L. J. AND DAVIS, C. L. (1914). The Effect of Alcohol on the Male Germ-cells studied by Means of Double Mating. *Sci.* 39, p. 476.

—— AND KIRKPATRICK, F. (1915). Sex Ratio in Pigeons together with Observations on the Laying, Incubation, and Hatching of the Eggs. *Bull.* No. 162, *R. I. Agric. Exp. Sta.* pp. 463–512.

COPEMAN, S. M. AND PARSONS, F. G. (1904). Observations on the Sex of Mice. *Proc. Roy. Soc. B*, **73**, pp. 32–48.

CREW, F. A. E. (1921). *Loc. cit.* (B).

—— (1923). *Loc. cit.* (B).

—— (1924). The Sex Ratio and the Question of its Control. *Inter. Rev. Agric.* **2**, pp. 554–570.

—— (1925). Prenatal Death in the Pig and its Effect upon the Sex Ratio. *Proc. Roy. Soc. Edinb.* **46**, pp. 9–14.

—— (1926). On Fertility in the Domestic Fowl. *Proc. Roy. Soc. Edinb.* **46**, pp. 230–238.

CUÉNOT, L. (1899). Sur la détermination du sexe chez les animaux. *Bull. sci. France et Belg.* **32**, pp. 462–535.

DARWIN, C. (1871). *The Descent of Man.* London.

DAVIS, W. H. (1918). Birth Statistics for the Birth Registration Area of the United States, 1916. *Bur. of Census*, Washington.

DAWSON, R. (1921). *The Causation of Sex in Man.* London.

DIGHTON, A. (1922). The Sex Ratio. *The Sport. Chronicle*, Sept. 23rd.

DONCASTER, L. (1914). *Loc. cit.* (B).

—— (1914). *Loc. cit.* (B).

—— AND MARSHALL, F. H. A. (1910). The Effect of One-sided Ovariotomy on the Sex of the Offspring. *Jour. Genet.* **1**, pp. 70–72.

DÜSING, K. (1884). Die Regulierung der Geschlechtsverhältnisse bei der Vermehrung der Menschen, Tiere und Pflanzen. *Jena. Zeit.* **17**, pp. 593–640.

ESSENBERG, J. M. (1923). *Loc. cit.* (B).

FEDERLEY, H. (1911). Vererbungsstudien an der Lepidopteren-Gattung Pygaera. *Arch. Rass. Ges. Biol.* **8**, pp. 281–338.

FIRCKS, VON, A. (1898). *Bevölkerungslehre und Bevölkerungsstatistik.* Leipzig.

FRANZ (1898). Zur Lehre des Aborts. *Beitr. z. Geburtsh.* **37**.

GINI, C. (1908). *Il sesso dal punto di vista statistico.* Palermo.

GOLDSCHMIDT, R. (1920). Untersuchungen über Intersexualität. *Zeit. indukt. Abst.* **23**, pp. 1–199.

—— (1923). *Loc. cit.*

GÖRTZ (1853). *Reise um die Welt in den Jahren 1844–47.* Stuttgart.

GÜNTHER, H. (1923). Lethaldispositionen und Sexualdispositionen. *Naturw. Korresp.* **1**, pp. 19–25, 46–52.

GUYER, M. F. (1909). *Loc. cit.* (B).

HALDANE, J. B. S. (1922). Sex-Ratio and Unisexual Sterility in Hybrid Animals. *Jour. Genet.* **12**, pp. 101–110.

HARRISON, J. W. H. (1916). Studies in the Hybrid Bistoninae. I. II. *Jour. Genet.* **6**, pp. 95–162, 269–314.

—— (1919). *Loc. cit.* (B).

—— (1919–20). Genetical Studies on Moths of the Geometrid Genus Oporabia (Oporinia), with a special consideration of Melanism in Lepidoptera. *Ibid.* **9**, pp. 195–280.

HARRISON J. W. H. (1920). The Inheritance of Melanism in the Genus *Tephrosia* (*Ectropia*) with some consideration of the Inconstancy of Unit Characters under Crossing. *Ibid*. **10**, pp. 61–85.

HAYS, F. A. (1921). Effect of Excessive Use of Male Rabbits in Breeding. *Breed. Gaz*. p. 66.

HEAPE, W. (1908). Note on the Proportions of the Sexes in Dogs. *Proc. Camb Phil. Soc*. **14**, pp. 121–151.

—— (1909). The Proportions of the Sexes produced by White and Coloured Peoples in Cuba. *Phil. Trans*. **200**, pp. 271–330.

HERTWIG, R. (1912). *Loc. cit*.

—— (1913).

HIRSCH, M. (1913). Ueber das Verhalten der Geschlechter. *Zentralb. f. Gynäk*. **37**, pp. 419–423.

HOFACKER, J. O. (1828). *Ueber die Eigenschaften welche sich bei Menschen und Tieren auf die Nachkommenschaft vererben, mit besonderer Rücksicht auf die Pferdezucht*. Tübingen.

HUXLEY, J. S. (1924). Sex Determination and Related Problems. *Med. Sci*. **10**, pp. 91–124.

IVANOV, E. (1913). Sur la fécondité de *Bison bonasus* × *Bos taurus*. *C. R. Soc. Biol*. **75**, pp. 376–378.

—— AND PHILIPCHENKO, J. (1916). Beschreibung von Hybriden zwischen Wisent und Hausrind. *Zeit. indukt. Abst*. **16**, pp. 1–48.

JACOBS, J. (1891). *Studies in Jewish Statistics, Social, Vital and Anthropometric*. London.

JASTRZEBSKI, DE, S. (1919). The Sex Ratio at Birth. *Eug. Rev*. **11**, pp. 7–16.

JENDRASSIK, E. (1911). Ueber die Frage des Knabengeburtenüberschuss und über andere Hereditätsprobleme. *Deut. med. Woch*. **38**, pp. 1729–1732.

JEWELL, F. H. (1921). Sex Ratios in Foetal Cattle. *Biol. Bull*. **41**, pp. 259–291.

JULL, M. A. (1923). Can Sex be Controlled? *Proc. Amer. Soc. Anim. Product*. pp. 92–98.

KING, H. D. (1918). The Effect of Inbreeding with Selection on the Sex Ratio of the Albino Rat. *Jour. Exp. Zool*. **27**, pp. 1–37.

—— (1921). A Comparative Study of the Birth Mortality in the Albino Rat and in Man. *Anat. Rec*. **20**, pp. 321–334.

—— AND STOTSENBURG, J. M. (1915). On the Normal Sex Ratio and the Size of Litter in the Albino Rat. *Ibid*. **9**, pp. 403–420.

KOHL (1859). Bemerkungen über die Bekehrung canadischer Indianer zum Christentum und einige Bekehrungsgeschichten. *In das Ausland*, **32**.

KÖRÖSY, J. (1898). *Die Sterblichkeit der Stadt Budapest in den Jahren 1886–1890*. Berlin.

KUSHAKEVITCH, S. (1910). *Loc. cit*. (A).

LAMBERT, W. V. AND KNOX, C. W. (1926). Genetic Studies in Poultry. *Biol. Bull*. **51**, pp. 225–236.

LENHOSSÉK, VON, M. (1903). *Das Problem der geschlechtsbestimmenden Ursachen.* Jena.

LEWIS, C. J. AND L. N. (1906). *Natality and Fecundity.* London.

LILLIE, F. R. (1917). *Loc. cit.* (B).

LITTLE, C. (1920). A Note on the Human Sex Ratio. *Proc. Nat. Acad. Sci.* **6**, pp. 250–253.

LYNCH, C. (1919). An Analysis of Certain Cases of Intra-specific Sterility. *Genet.* **4**, pp. 501–553.

MACDOWELL, E. C. AND LORD, E. M. (1926). The Relative Variability of Male and Female Mouse Embryos. *Amer. Jour. Anat.* **37**, pp. 127–140.

MACHENS, A. (1915). Fruchtbarkeit und Geschlechtsverhältniss beim veredelten Landschwein. *Berl. tierärzt. Woch.* **31**, pp. 559–562.

MALINS, E. (1903). Some Aspects of the Economic and of the Antenatal Waste of Life in Nature and Civilisation. *Jour. Obstet. Gynaec.* **3**, pp. 307–319.

MOHR, O. L. (1923). Modifications of the Sex-Ratio through a Sex-linked Semi-lethal in *Drosophila melanogaster. Studia Mendel.* pp. 266–287.

MONTGOMERY, T. H. (1908). The Sex-Ratio and Cocooning Habits of an Aranead and the Genesis of Sex-Ratios. *Jour. Exp. Zool.* **5**, pp. 429–453.

MORGAN, T. H. (1911). An Alteration of the Sex-Ratio induced by Hybridisation. *Proc. Soc. Exp. Biol. Med.* **8**, pp. 82–83.

—— (1914). Two Sex-linked Lethal Factors in *Drosophila* and their Influence on the Sex-Ratio. *Jour. Exp. Zool.* **17**, pp. 81–122.

MRŠIĆ, W. (1923). *Loc. cit.* (B).

NICHOLS, J. B. (1907). The Numerical Proportion of the Sexes at Birth. *Mem. Amer. Anthrop. Ass.* **1**, pp. 249–300.

PARKER, G. AND BULLARD, C. (1913). On the Size of the Litters and the Number of Nipples in Swine. *Proc. Amer. Acad. Arts. Sci.* **49**, pp. 399–426.

PARKES, A. S. (1923). Studies on the Sex-Ratio and the Related Phenomena. I. Foetal Retrogression in Mice. *Proc. Roy. Soc.* B, **95**, pp. 551–558.

—— (1923). III. The Influence of the Size of the Litter. *Ann. Appl. Biol.* **10**, pp. 287–292.

—— (1923). IV. The Frequencies of Sex Combinations in Pig Litters. *Biomet.* **15**, pp. 378–381.

—— (1924). II. The Influence of the Age of Mother on the Sex Ratio in Man. *Jour. Genet.* **14**, pp. 39–47.

—— (1924). V. The Sex-Ratio in Mice and its Variation. *Brit. Jour. Exp. Biol.* **1**, pp. 323–334.

—— (1925). VI. The Effect of Polygyny. *Ann. Appl. Biol.* **12**, pp. 211–217.

—— (1925). VII. The Effects on Fertility and the Sex-Ratio of Sub-Sterility Exposures to X-rays. *Proc. Roy. Soc.* B, **98**, pp. 415–436

PARKES, A. S. (1926). Proportions of the Sexes in Man. *Eug. Rev.* 17, pp. 275–293.
—— (1926). The Mammalian Sex Ratio. *Biol. Rev.* 2, pp. 1–51.
PEARL, R. (1913). On the Correlation between the Number of Mammae of the Dam and Size of Litter in Mammals. *Proc. Soc. Exp. Biol. Med.* 11, pp. 27–32.
—— (1917). Effect of Heavy Laying in the Fowl. *Sci.* 46, pp. 220–222.
—— AND PARSHLEY, H. M. (1913). Data on Sex-Determination in Cattle. *Biol. Bull.* 24, pp. 205–225.
—— AND PEARL, M. (1908). On the Relation of Race Crossing to Sex-Ratio. *Biol. Bull.* 15, pp. 194–205.
PEARSON, K. (1908). *The Chances of Death, and other studies in Evolution.* London.
PINARD, A. AND MAGNAN, A. (1913). Sur la fragilité du sexe mâle. *C. R. Acad. Sci.* 156, pp. 401–403.
POWERS (1877). *Tribes of California.* Washington.
PRIESTLEY, W. O. (1887). Pathology of Intrauterine Death. *Brit. Med. Jour.* pt 1, pp. 660–669.
PUNNETT, R. C. (1903). *Loc. cit.* (A).
QUETELET, A. (1872). *Tables de mortalité et leur développement d'après le plan d'une statistique internationale et comparée.* Bruxelles.
RAUBER, A. (1900). *Der Ueberschuss an Knabengeburten und seine biologische Bedeutung.* Leipzig.
RAWLS, E. (1913). Sex-Ratios in *Drosophila ampelophila. Biol. Bull.* 24, pp. 115–124.
RIDDLE, O. (1916). *Loc. cit.* (B).
—— (1924). *Loc. cit.* (B).
ROUTH, A. (1914). Ante-natal Hygiene: its Influence upon Infantile Mortality. *Brit. Med. Jour.* pt 1, pp. 355–363.
RUST, J. L. F. (1902). *Das Geschlecht der Fehl- und Totgeburten.* Dissertation, Strassburg.
SADLER, M. TH. (1830). *The Law of Population.* London.
SANDERSON, J. (1879). Polygamous Marriages among the Kaffirs of Natal and Countries around. *Jour. Anthrop. Inst.* 8, pp. 254–260.
SCHLEIP, W. (1913). Geschlechtsbestimmende Ursachen im Tierreich. *Ergeb. u. Fortschr. d. Zool.* 3, pp. 165–238.
SCHRADER, F. (1923). The Sex-Ratio and Oogenesis of Pseudococcus. *Zeit. indukt. Abst.* 30, pp. 163–182.
SCHULTZ, A. H. (1918). Studies in the Sex Ratio in Man. *Biol. Bull.* 34, pp. 257–275.
SCHULTZE, O. (1903). Zur Frage von geschlechtsbildenden Ursachen. *Arch. mikr. Anat.* 63, pp. 197–257.
SEILER, J. (1920). Geschlechtschromosomenuntersuchungen an Psychiden. *Arch. f. Zellf.* 15, pp. 249–268.
SMITH, G. AND HAIG-THOMAS, R. (1913). On Sterile and Hybrid Pheasants. *Jour. Genet.* 3, pp. 39–52.
STARKWEATHER (1883). *The Law of Sex.* London.

STEWART, C. H. (1910–11). The Sex and Age Incidence of Mortality from Pulmonary Tuberculosis in Scotland and in its Group of Registration Districts since 1861. *Proc. Roy. Soc. Edinb.* **31**, pp. 352–373.

STURTEVANT, A. H. (1920). *Loc. cit.* (A).

SUCHETET, A. (1897). Problèmes hybridologiques. *Jour. d'anat. et physiol.* **33**, pp. 326–355.

SUMNER, F. B., McDANIEL, M. E. AND HUESTIS, R. H. (1922). A Study of Influences which May Affect the Sex Ratio of the Deer-Mouse. *Biol. Bull.* **43**, pp. 123–165.

TAUSSIG, F. J. (1910). *Prevention and Treatment of Abortion.* St Louis.

THOMAS, W. I. (1913). *Report on Ibo-Speaking People of Nigeria.*

THURY, M. (1863). *Ueber das Gesetz der Erzeugung der Geschlechter.* Leipzig.

TROUESSART, E. (1907). Hybrides de paon et de poule cochinchinoise. *L'Acclimat.* No. 103, p. 645.

WAITZ, TH. (1859). *Anthropologie der Naturvölker.* Leipzig.

WENTWORTH, E. N. (1914). Sex in Multiple Birth. *Sci.* **39**, p. 611.

WHITEHEAD (1914). Quoted by Routh.

WHITMAN, C. AND RIDDLE, O. (1919). Orthogenetic Evolution in Pigeons. *Carn. Inst. Publ.* No. 257.

WILCKENS, M. (1866). Untersuchungen über das Geschlechtsverhältniss und die Ursachen der Geschlechtsbildung bei Haustieren. *Landw. Jahrb.* **15.**

WILLIAMS, C. B. (1917). Some Problems of Sex-Ratios and Parthenogenesis. *Jour. Genet.* **6**, pp. 255–268.

WILLIAMS, J. W. (1923). *Obstetrics.* London.

WITSCHI, E. (1922). Chromosomen und Geschlecht bei *Rana temporaria*. *Zeit. indukt. Abst.* **27**, pp. 243–255.

AUTHORS' INDEX

SUBJECT INDEX

Printed in the United States
By Bookmasters